JACK KLAFF

Published October 2018

A CIP Catalogue record for this book is available from the British Library.

ISBN: 978 1 78521 557 5 (print)
978 1 78521 611 4 (eBook)

Library of Congress control no. 2018950619

Published by Haynes Publishing,
Sparkford, Yeovil, Somerset BA22 7JJ
Tel: 01963 440635
Int. tel: +44 1963 440635
Website: www.haynes.com

Printed in Malaysia.

Series Editor: David Allsop.
Front cover illustration by Alan Capel.

CONTENTS

Nothing conveys the impression of substantial intellect so much as even the sketchiest knowledge of quantum physics.

LET'S GET QUANTUM PHYSICAL

Whether you are trying to explore the universe, which is very, very big, or the 'quantum realm', in which particles of light and matter are very, very small, nothing can be stated with clearer conviction than this: nobody understands what's going on. The world's greatest physicists have openly admitted this for more than 100 years. So, if you've ever lamented your own incomprehension, let yourself off the hook immediately. Welcome to a bluffer's paradise, where no one need be afraid, nor should anyone feel stupid.

Of course, where physicists do have knowledge and understanding, prudence dictates that astute bluffers gain some appreciation of it. This will not be too difficult. The universe is, after all, very large, and its smallest entities are extremely small – and their conduct is unusual, to say the least.

Happily, such an appreciation may be obtained

without formulae, equations or fractions. It may seem to be taking the word 'appreciation' too far, but you are going to have to feign, find or affirm some sense of wonder. Nature is full of marvels at the best of times. In the quantum realm, as the great Danish physicist, Niels Bohr, used to say: if people didn't find these phenomena baffling, wild and shocking they weren't taking it in. At the end of this book, you might agree that Bohr was understating things a bit.

Inevitably, overwhelming questions will keep coming up. Nowadays such questions can neatly be batted away with the words: 'Yes, well, of course they're looking into that at Cern.' You hint at deeper knowledge by saying that Cern is an acronym for Conseil Européen pour la Recherche Nucléaire and by being specific about the actual location of the famous 'collider' – near Geneva, beneath the Franco-Swiss border. (Not too far away from Goldfinger's lair in the eponymous Bond book, actually.)

Mentioning Cern will become an automatic response for you, and one that falls somewhere between 'They deal with that in accounts', 'My mother-in-law will know' and 'Give the ball to Brian,' No one should ever tire of saying, 'Yes, they're addressing that at Cern.'

Nothing conveys the impression of substantial intellect so much as even the sketchiest knowledge of quantum physics, and since the sketchiest knowledge is all anyone will ever have, never be shy of holding forth with bags of authority about subatomic particles and the quantum realm without having done any science whatsoever.

After all, what else is the act of bluffing about?

It won't do any harm, however, to have a tangential grasp of your new area of expertise. The word 'quantum', as you will see, refers to tiny entities, originally units of energy. A quantum is an amount – the word has the same root as 'quantity'.

And 'mechanics' is the branch of mathematics which deals with motion and the forces which produce motion. 'Quantum mechanics' has absolutely nothing grease monkey-ish about it. That's it in a nutshell.

This book sets out to guide you through the main danger zones encountered in discussions about the quantum universe. It also equips the dedicated bluffer with a vocabulary and a range of evasive techniques that will minimise the risk of being rumbled. It will lend you a few easy-to-learn hints and methods that will allow you to be accepted as a physics expert of rare ability and experience. But it will do more. It will give you the tools to impress legions of marvelling listeners with your knowledge and insight – without anyone discovering that before reading it you didn't know the difference between a boson and a fermion or a meson, or come to that a photon, pion or gluon.

However tempting, try to avoid anything like 'dreamon', carryon, whatson, or admit that you're 'puttingiton'. And try not to misspell or mispronounce 'hadron'.

Bluffers are smarter than this (especially when they're engaged in a good bluffon).

Meanwhile, the boffins are carrying on their work, largely unknown and uncelebrated (apart from the odd Nobel prize).

Physicists, like all scientists, are supposed to do

their work, discover natural laws and give us an ever more profound understanding of the universe around us. They are not supposed to make anything practical at all. Scientists are not technologists. They are people for whom application is a dirty word. And yet. And yet. . . 'The discoveries that scientists have made in the quantum realm have utterly transformed the way in which life, the universe and everything are thought about by physicists. But what exactly have their dabblings in the quantum universe done for us?'

It's the kind of question you might find on the Internet, which itself wouldn't exist without quantum mechanics.

And that's apart from the entire computer industry, desktops, laptops, tablets, smart phones, everything to do with IT, televisions, sound systems, so many household goods, so many toys, digital watches, lasers, MRI scanners and fibre-optic cables for telecommunications... not to mention the atom bomb, every nuclear weapon since, and every smart bomb too.

So, yes, it's a good question. What exactly has quantum mechanics done for us?

Readon.

PREPARING YOURSELF

DOES SIZE MATTER?

Whichever way you approach discussions about the universe and the quantum realm, the notion of scale will somehow always be present. But remember that it should be beneath you to be impressed by size; it is for other people to reel and gasp. Scale is something with which you must appear to be comfortable and familiar.

Physicists naturally possess what bluffers must somehow acquire: the ability to drop phenomenally massive numbers into the conversation with ease, airily disparaging 'gee-whizzery' while trotting out awe-inspiring facts and figures.

Where the tendency is towards 'big', you need to appreciate that a light year measures the distance you could cover in a Julian year (that's a normal calendar year of 365.25 days), zooming at 186,283 miles per second. The nearest galaxy to Earth, newly discovered, is a trillion miles away – remarkably near as galaxies go, but still pretty far, even for a commercial traveller. (And if you do go, don't mention Brexit.)

At the other extreme, you need to have some idea of what is meant by 'small'. All matter can be broken down into atoms. Atoms are small. They are smaller than affordable apartments in Manhattan, they are smaller than one of Heston Blumenthal's main courses, they are even smaller than the chance that a politician will give you an honest answer. The full stop at the end of this sentence will be a tiny blob of ink about a quarter of a millimetre wide which will contain close to four billion atoms.

Having taken that into account, a reminder is necessary. To a human being, or to a full stop, an atom is small. But to a subatomic particle, any one of those four billion atoms in that full stop is enormous.

At the centre of an atom is a nucleus – itself made of particles. The nucleus feels within the atom as a regulation red ball might feel in a cricket ground. The electrons that surround the atom would be akin to peanuts circling at the outskirts of the ground's car park. There are no players, spectators or refreshment-sellers around. Atoms are usually described as being made up of 99.9999999999999% 'empty space'. Without that empty space, the entire human race would fit into a space the size of a sugar cube.

Pointing to such a cube, or mentioning how many atoms there are in a grain of coffee, or Parmesan cheese, or beach sand, or, best of all, in a pretty freckle on human skin, may be a game well worth the candle. But a confident bluffer can take it further.

With the concept of 'touch' of any kind – innocent,

compromising or downright culpable – the physics of today provides a most helpful get-out clause. No atoms of any one object can ever touch the atoms of another.

For hundreds of years, people have talked of great performers having magnetism, or public moments having electricity, but it has taken humanity more than a few millennia to realise how much electromagnetism there exists within every atom and therefore within everything on Earth. Electromagnetism is not regarded as a strong force within the atom. But even though the 'empty space' within atoms may be devoid of mass – of matter, of 'stuff' – the electromagnetic charge within each atom is powerful indeed. How powerful? Why don't people fall through the ground as they walk through a town or city? Because the negative electromagnetic charge on the outside of the atoms in a pavement exerts a repulsive force on the negative charge of the atoms in the soles of a shoe. That is the same repulsive force in action when hammers hit nails, boxing gloves punch heads or rackets thwack tennis balls.

The charges around the atom create the illusion and even the feeling of touch. Still, however hard they are pushed, the force surrounding atoms far exceeds even the most powerful bar magnets, so atoms cannot be made to come into contact with each other. There is always that layer of separation, even around scissors and hair, food as it's digested and, therefore, between two pairs of lips.

Whatever the status of human relationships, lovers have no option but to accept the old adage: 'Let there be space in our togetherness.'

THE PARADOX OF THE BELIEF SYSTEM

Few forms of knowledge are so closely interwoven with human belief systems as studies concerning the quantum universe.

When discussing the very tiniest particles of light and matter, or wrestling with the very big universe, your belief system, and the belief systems of the people whom you will be impressing, are going to be powerfully challenged.

Because you are dealing with fundamental entities in nature, it will be impossible to avoid disturbing and challenging the most basic beliefs of those around the table. The mere mention of a particle will raise the issue of consciousness in snails, and how this relates to God.

Everything that anyone has ever held dear, every blind assumption, every hard-won prejudice may be horribly threatened. Clearly, this sort of thing should not be allowed; and, certainly, no one else should have to witness it.

To prepare yourself for 'deipnosophy' (the noble art of excelling at dinner parties), you need to know where you stand. Behind closed doors you will have resolved all those profound questions which, however slightly, might have upset your own world view. You will have absorbed this knowledge – and been shaken and stirred – but you will have managed to keep your cherished notions intact. Before you emerge again to re-engage and mingle with your friends, you will have discovered how to use the very same knowledge to provide you with an opposite, more pleasing interpretation.

It will involve no effort at all to find some subatomic occurrence that has a bearing on religion, or politics, or marriage or morals, or art or azaleas, or music or marmalade, or snooker or sex. Each precious phenomenon must be seized on, every analogy squeezed until it squeaks.

TWO TYPES OF PEOPLE

It has been said that there are two types of people: those who divide the world into two types of people and those who don't.

Well, having said that, even in the restricted field of explaining quantum mechanics – and finding those rare analogies that are convincing and accurate – there are two types of people. These were identified by physicists just after the war when the American theoretician Julian Schwinger was in his own way cock of the walk. As the British-born US-based physicist Freeman Dyson told the story, quantum theorists used to tell their colleagues 'How to do it.' Schwinger used to present his work as if he were saying, 'Only I can do it'.

Those two approaches ought to be considered by any bluffer. Do you seek to make things crystal clear as you blag your way through? Or do you indicate that you and only you are the only person who understands it all?

Do you go for a combo, for both?

No harm in spreading the knowledge, it's good knowledge to spread, but the whole practice of bluffing would be wasted if you didn't have a little touch of Schwinger in the night.

While we're talking about dividing people into types, there are trillions of ways whereby ideas about quantum mechanics can divide people in trillions of ways, but it's instructive to identify two distinct approaches whereby those who've encountered quantum physics will choose to interpret this relatively new science, two views of the universe if you like, two schools of thought. Here's how you can see them coming.

1. The Holists
Members of this school are, broadly speaking, holistic. Holists are interested in the whole picture, except when it comes to the letter 'w'. As time goes on, more and more holists are likely to come out of the (w)ood(w)ork.

2. The Reductionists
Those who belong to this school are interested in the parts of the picture. All they want to know is 'How small?', 'How many?', 'How wide?', etc., as if the quantum universe was a watch that could be taken apart to the smallest screw to find how it all works.

Holists are generally interested in quality. Reductionists are interested in quantity.

Holists, to paraphrase the poet William Blake, can see the universe and eternity in an atom – or even a grain of sand. They have likened the universe to an enormous ball of string. No one can see its end or its beginning, yet if you draw on the string at any part of the ball, the ball becomes tighter throughout. Try to pull one strand of it, and all of it changes. Holists say that this involves a deeper reality than mere interrelationship. For them, wherever you tweak the string, there is the entire universe because all of it changes. Everything is part of everything, every part is part of every part. There is oneness and only oneness. If it sounds like religion, it depends on how you define religion.

Reductionists concentrate on the individual atom without relating it to anything else. They think you should chop up the string ball and measure the bits.

Holists think the scientific discoveries of the twentieth century should encourage us to rethink everything. This is sometimes called a 'paradigm shift' – the altering of a model, in this case the whole model. They want to unite, to bind, to join. Reductionists don't. They like the word 'discrete', meaning 'separate from everything else'.

Modern physics, like love, or football, or being a teenager, defies language.

Reductionists think it is unnecessary to make so much of the fact that the iron at the core of planets is also contained in human haemoglobin. Holists think it's poetic and meaningful that human beings have stardust in their blood.

Holists find coincidences intensely moving; reductionists demonstrate why they aren't coincidental.

Holists accept a certain floppiness in their ideas; reductionists insist on rigour.

Holists quite like the circuitous approach. A reductionist will even do a cutting gesture while explaining that space itself is curved so you will, at least, get that *straight*.

THE LANGUAGE BARRIER

When you come up against the limits of the quantum universe, and indeed the limits of your own understanding, you will inevitably come up against the extremes of what you can express linguistically, so to speak.

For instance, the notion that in the beginning everything was in one place and, lo, that place was nothing. No thing. It was a nothingness, so nothing like its nothingness took nothingness to untold extremes of nothingness. (Modern physics, like love, or football, or being a teenager, defies language.)

It has never been easy for anyone to find words to explain the deep mysteries of the universe. How to explain new phenomena? How to describe them? How to tell people about entities and occurrences which are so far removed from anything hitherto known about or understood that the brain starts doing backflips and wants to join the circus.

The difficulties begin with names. You will already be expecting to encounter many words in the quantum realm which end in '-on'. As you know, there are hadrons, which may be baryons or mesons. Mesons may be pions or kaons, while muons and tauons are leptons, not to be confused with sleptons – and so on.

You may be tempted to sort them out. Don't. At the University of California at Berkeley, in an almanac which records new particles, the list was already running to more than 2,000 pages in the 1990s. The Italian physicist Enrico Fermi said if he could remember all the names given to particles he would have been a botanist, and if such things really do interest you, you need to get out more. Besides, as early as 1964, the American physicist Murray Gell-Mann and others have felt that all those differently named entities were actually combinations of a relatively few truly elementary particles. And that should be your suggestion too. Probably. It is small wonder, perhaps, that the Nobel Prize-winning Gell-Mann is these days devoting himself more to the study of language than to physics.

Niels Bohr, who used to have days of intense discussion about almost every word he used, was called a bad lecturer, but often he just wanted to get the language right. He once paused for a long moment before saying, '…and', then took another very long pause before he said, '…but'. (It was he who reportedly gave a baffled French ambassador the happy and heartfelt greeting: *'Aujourd'hui!'*) At least he conceded the problem. It took him several weeks of hard thought and discussion to come up with this: 'We are somehow suspended in language.'

You could do no better yourself.

Blaming language is your parachute, your cavalry, your safety net. A reminder about linguistic difficulties will send sharp questions ricocheting back to their posers; and if or when you have dropped yourself in it, talking about the lack of suitable words can soften the horror-movie eyes around the table.

The language used by physicists is mathematics – though many of them need help with it from real mathematicians. Their best-known trick is to shut their eyes and find some polysyllabic blah-de-blah-blah to make it seem deep and meaningful. Physics is chock-full of words which have been poached and which may once have had a sporting chance of meaning something. Most could not be more confusing if they were designed to be so, and the suspicion is that they were.

It may also be a useful gambit to cite languages other than your own, such as Algonkian spoken by the Blackfoot people of Alberta, Canada, which emphasises change and movement in nature, rather than finding names for things. For example, instead of saying, 'I saw someone long ago,' they say, 'I saw him/her in the far away.' This so impressed the eminent American scientist David Bohm that he proposed that an entirely new kind of language should be developed for the quantum universe. An English one, of course.

VISUALISING

A few words can trigger pictures and movies in your mind.

Stop them. Make sure the bouncers at the entrance to your brain are ready to impose the strictest vetting standards ever. Bad image; wrong image; misleading image. They ain't coming in.

As early as 1926 the great physicist Schrödinger asked the readers of one of his papers not to visualise the phenomena he was describing. They should just understand the mathematics, which was convenient. And that warning has been repeated countless times down the ages. A physics textbook by Halliday, Resnick, and Walker, published in 1993, states that: 'The single electron *does* interfere with itself. But don't try to visualise how it does so.'

DO: Be patient. You'll soon find out in detail how electrons interfere with themselves.
DON'T: Overdo the explanation of why this is funny.

Of course, unbidden or forbidden pictures will pop into our minds. They are ungovernable.

Wait. Is that true?

Does that happen to everyone?

Are there members of our species who have ideas and who don't have any mental pictures in their minds about those ideas? Is their visualising deferred? Do they need to get the maths right before they so much as dare to take an imaginary peek? Maybe, just maybe there are such people.

When Schrödinger issued that warning against visualising, the great physicist Heisenberg wrote to the great physicist Pauli: 'What Schrödinger writes about visualisation makes scarcely any sense, in other words I think it is crap.'

It is just possible that Heisenberg didn't visualise. That he couldn't even visualise how anyone needed to visualise. That his mind didn't work that way. That because Schrödinger's squiggles made sense, Nature had to measure up, or be rejected, or at least resit the exam.

Not that long ago on the Internet an eminent computer scientist called Caterina wrote a public thank you note to her maths teacher in Italy. When she was a child Caterina hated those pictures of cakes and pies which she was shown in her arithmetic classes. Luckily for her she encountered a maths teacher who told her that mathematics is abstract. She just needed to understand the concepts and the logic of the figures on the symbols and how they related to each other. She didn't need the cakes and the pies.

Beware of suggesting that physicists have no visual sense. Of course it's true that biologists show videos that are feasts for the eyes while physicists use state-of-the-art projectors to show dull transparencies covered with scrawled figures and letters that are almost all Greek. And, yes, the physics buildings on university campuses are often ugly. And, all right, there may be some physicists who aren't exactly fashion-conscious.

But let's be fair. Schrödinger must have been a visualiser. Another Nobel-winning physicist, Richard Feynman, frequently used diagrams, mainly thought in pictures and always made sure that he could see what an abstract theory meant in the visible world. Feynman's great contemporary, Murray Gell-Mann publicly castigated physicists who spoke too much about the beauty of their equations and didn't relate a beautiful theory to the real and beautiful world.

You will make a choice as to whether you do say or don't say these things.

But in electricity the terms 'current' and 'flow' don't really describe what electricity does. Benjamin Franklin coined those terms. Could he have anticipated that people would see electrical wires as streams, rivers or hosepipes? And might worry about leaks from plug sockets. Franklin was also responsible for making an arbitrary choice. A 'negative' charge is not depressed, nor in debt, nor unlucky in love, nor compulsively critical. A 'positive' charge isn't contented or upbeat or supportive. He just named them the way some people name goldfish – and others name particles. As it happens, the fact that an 'electron' has a 'negative' charge might not be a problem for you but the fact that they aren't 'positive' is something of a pain for physicists (which might be poetic justice).

You will genuinely hear physicists say that the 'universe is flat'. They're talking about Euclidean geometry and its planes. When they look up (from their pages or screens) they can see that the universe is not literally flat. Surely. But that doesn't stop them from using words like 'trampoline' or 'fabric' when talking about four-dimensional space-time. Something with perceivable length, breadth and depth – like 'a cube of jelly' – would convey the idea in one second. Flat.

'Quarks' were so named because the physicist Murray Gell-Mann was quoting a line from *Finnegans Wake* by James Joyce: 'Three quarks for Muster Mark.' Understanding Mark's relevance to the whole thing needs more enthusiasm than anyone can Muster. Quarks, like sailors, were seen to go around in threes.

In 2013 a particle containing a fourth quark was confirmed…

In 1897 the English boffin JJ Thomson discovered that there could be particles smaller than atoms, and described them as 'plums in a pudding'. Or maybe it was 'raisins in a bun'. Then in 1913 the great Danish theorist Niels Bohr depicted electrons in orbit around the nucleus at the centre of the atom.

Although Bohr's idea of orbiting mini-planets wasn't quite right – as he himself admitted, experiments suggested that particles might 'spin'. But it's now clear that it's misleading to conjure up the image of a small spinning object or to think of a particle as a ball. When certain elementary particles move through a magnetic field they are deflected. That creates the illusion of spin. But the particles are more like insubstantial little magnets. Magnets poised at an angle which never changes. Their orientation remains the same. They are not, repeat not, at that level, solid objects.

DON'T SAY (OR SING):
Do a particle orbit?
Do a particle 'spin'?
Don't know where it's goin'.
Don't know where it's bin.

DO SAY: *They're looking into that at Cern.*

When science explainers say that subatomic particles and atoms are the 'building blocks' of Nature – and the vast majority of them do – they are just being silly, and not listening to themselves. Nothing, nothing at all, could be less like a building block than an atom.

TACTICS FOR TIGHT CORNERS

Whenever the universe becomes a topic of conversation, certain questions will always be hovering around the edges – if there are any edges (and that's one of the questions). These include:

- How did the universe begin?
- Who cares?
- If the universe had a beginning, will it have an end? Or a new series in the autumn?
- What has 'quantum' got to do with it? And will it affect the Dow-Jones Index?
- I sort of get it, but just when I do, I lose it again. It's as if it's in the cracks of my understanding. Is that where it should be?
- Can I have a Bloody Mary?

Some of these penetrating queries may seem to be unanswerable. And that is because they are. But defeatism is alien to science as well as to bluffing. And since the scientists who are bold enough to deal with these questions are so often dealing in conjecture – or worse – you may as well at times follow suit. Allow the bluffing muse to descend and announce: 'Physicists now believe that before the universe began, everything was a substance not dissimilar to bubblegum'; or, 'Of course you can't see them but there aren't just three dimensions of space and one of time, there are more, maybe 10 or 11 or 12, or, yes, hundreds, millions, billions and billions of dimensions'; or, 'Last week, cosmologists declared that the universe was stacked between two bookends, now they're saying it's shaped like the upside-down torso of a man in boxer shorts.'

These are not far from actual quotes and are not altogether made up. But remember to balance conviction with caution.

Never commit yourself about the outer limits of the universe or the quantum realm even with a 'probably'. Anything you utter with certainty, or declare to be 'probably true', could return to haunt you and, it can be said with confidence, probably will. If you know what's good for you, a 'possibly' is the farthest you will go.

To avoid being thought a delinquent in the light of tomorrow's new data, you need to make the lack of language work for you by:

a) saying that we are up against the limits of it;

b) frequently using the word 'possibly' as an insurance policy; and

c) capping all statements about problems that have not been solved and theories that have been proved unassailable, or the finding of the smallest particle in the universe, with the caveat 'so far'.

With some of the more difficult questions, a gentle attack may be a good defence. The 'dismissal' tactic is ideal since it conveys the impression of very deep and thorough knowledge and prevents any deviation from your chosen conversational course. The dismissal is best exemplified by the simple but salutary sentence: 'The question doesn't apply.'

For decades scientists said that there was no before. That is to say: no 'before' before the Big Bang. As has already been mentioned, everything was in one spot that wasn't even a spot, it was less than zilch and then that less-than-zilch went 'Vrooom' and the universe just 'grew'.

How could they know that? Yes, they interpreted it from the mathematics.

And what was there before the 'Vroom'? The answer was as cold as a padlock: 'the question doesn't apply'. Everyone who was not a physicist was at that time hectored into not thinking of a 'before'. The question didn't apply. Even so much as considering a 'before' had to be avoided because there was nothing there, indeed there was no 'there' there, just as there was no 'because' because 'because' needed a 'dum-de-dum' in order to have gone from 'dum-de-dum' to 'dum-de-dee' because of 'blah-de-blah'. Which didn't happen, nor did 'happen' happen, nor did 'there', nor did 'anywhere'.

Well, time passed and slowly but surely a substantial number of influential physicists came to accept what everyone else could see was blindingly obvious.

The latest thinking among serious physicists is that something must have gone on that PRECEDED the BIG BANG.

The universe, after all, had to have had a BEFORE.

There was no apology from physicists. There was not even an acknowledgement. Not one of them came out and said, 'Your splutterings were perfectly reasonable. We patronised you, in fact we were really rather rude, but of course we were talking nonsense. So, sorry. You were right. We just couldn't visualise what you were visualising; we couldn't even visualise that you were visualising.'

How do scientists get away with such high-handed behaviour? The question doesn't apply.

THE QUANTUM REALM

The study of subatomic particles is called quantum mechanics. This is strange because the word 'quantum' is Latin for 'how much' or 'how great'. In this case, not a lot. And you can see within a nanosecond (a very small fraction of a second) that it is entirely alien to this subject to talk of anything 'mechanical', 'mechanistic' or 'machine-like'.

The term 'quantum' was coined in 1900 by the great physicist, Max Planck. The quantum unit was to represent a 'package of energy'. Despite the temptation to add 'on' to the end of any word to reinforce one's quantum credentials, one can only hope that neither the prolific namers of particles nor naff punsters will ever mention anything involving 'Planckton', which really won't do. Indeed, bluffers should avoid all such clumsy jokes, especially the all-too-obvious 'Thick as a Planck'. Besides which, Planck was exceptionally bright. (The same restrictions on gags applies to 'Thundering' Bohr, 'Hugh' Schwinger and 'Atomic' Bohm.)

THE QUANTUM MOMENT

Max Planck's professor advised him not to do physics because it was thought at that time that all its problems had been solved. Planck ignored his advice. In 1900 he found a problem to solve and, in solving it, set a million more in motion.

The problem bore the name, unfairly perhaps, of 'Violet Catastrophe'. You can allow a few sniggers before bringing everyone to order by adding that it is also known as the 'Ultraviolet Catastrophe'. It occurred as a result of 'black body radiation'. A black body – a black object – tends to absorb radiation when cool; but if you heat it sufficiently, it will emit radiation by going red hot. The hotter it gets, the farther up the spectrum the glow issuing from it will get, through an orange and a blue heat, all the way to ultraviolet (the sort of light that picked out teeth and specks on a jacket in the good-old golden days of the discotheque).

For a physicist, seeing something work is not the point; the point is that the mathematics of it must work. The sums identified by Planck implied that the progress through the spectrum of the differently coloured glows would be shattered in a burst of catastrophic ultraviolet radiation. But however intense the heat, no such destructive explosion occurred. Something was wrong with the calculations. So Planck did the obvious thing. He cheated. He squeezed a number into the sums: 0.000 000 000 000 000 000 000 000 00066.

Then he attempted to do that most difficult thing for a scientist: he went back to real life. But whatever he tried, he could not get rid of that embarrassing little subterfuge.

It was some time before he realised that the cheat was the whole point. The cheat was Nature's cheat. The black body did not heat, or cool, in one smooth run of energy but in little jerks or spurts. (On reflection, perhaps Quantum is a better name than Jerk or Spurt – but the question doesn't apply.)

You should call that number, with its 27 zeros before the 66, Planck's constant.

In physics, a 'constant' is a quantity that is universal in nature and constant in time. In other words, a 'constant' occurs always and everywhere in nature. Newton's gravity and the speed of light used by Einstein are constants. Planck's constant is in that exclusive club.

For years, Planck's discovery was not taken by other physicists to be a revolution. Then when it was, it was precisely the kind of revolution he fought for the rest of his life to prove it wasn't. He had created a monster which threatened the very basis of classical science. Again and again he returned to the problem he had already solved, even making ready to undo his own work and diminish his ranking as a physicist. To remark that there never was a more reluctant springboard than Planck is a truly unintentional breach of the no-pun rule.

THE MEDIUM AND THE MESSAGE

Quantum physics started off being concerned with the composition of matter, the properties of light, and the interaction between light and matter. In 1900, when

Planck was playing with fire and tinkering with his maths, what he was examining was the interaction between light and matter. The young Albert Einstein explained in 1905 how photons – particles of light – could be directed at a metal and dislodge electrons – particles of matter.

Bonus bluffing marks will come your way when you declare that Einstein's examinations of the photoelectric effect were indispensable when TV was invented.

Einstein won his Nobel Prize not for the general theory of relativity, nor for his big hair, extravagant moustache or his habit of not wearing socks, but for his work on the photoelectric effect.

(Note that no one has won a Nobel Prize for anything to do with relativity. It is not seen as having been properly tested. Scores of Nobels have been awarded for quantum physics.)

Light is just one of a number of agents of transmission which dislodge, illuminate, attract, repel, bind or explode all the materials of nature in a constant dance. You may remember that you will receive much gratitude and admiration for imparting the simple information that there are, basically, two main categories whereby subatomic particles may be classified:

1. particles of matter; and
2. messenger particles.

The suspicion is that for each kind of matter particle there exists a 'symmetrical' messenger particle. If so, it would create a marvellous balance of medium and message throughout the universe which you should call 'supersymmetry'. This is a fantastic word for bluffers and it can be used to great effect whenever discussing the notion of 'balance' in a universal context.

It will probably do you no lasting harm to know that particles of matter have the generic name, fermions, and all messenger particles, whatever else they are called, are also known as bosons.

Saying that fermions are scared of commitment can be a delectable conversational gambit in mixed company.

It is worth gaining some extra bluffing points by mentioning the 1946 talk given by the British-born physicist Paul AM Dirac. Dirac was a taciturn gent, but this lecture was noteworthy for other reasons, for in the course of it he suggested the names 'fermion' and 'boson', paying tribute to two esteemed scientific colleagues. Enrico Fermi was most unusual in that he was an excellent experimentalist as well as a top-class theoretician. Satyendra Nath Bose understood light and matter so profoundly that Einstein himself headhunted him for collaborations. Much came out of their work together, not least the hologram.

Fermions get their mass from a unique particle, the now legendary Higgs boson; quite rightly, somewhere along the way, science writers ceased to deploy Higgs's apostrophe.

Saying that fermions are scared of commitment, individualistic and selfish, and that bosons are the outgoing, surrendering relationship particles, can be a delectable conversational gambit in mixed company.

QUANTUM MECHANICS

You can divide the early history of the development of quantum mechanics into three main periods. During the first, the discoveries of Planck and Einstein were demonstrated but not fully understood. During the second, Niels Bohr's theory of 1913 was explored; his work with atoms and light was enormously helpful without, strictly speaking, being comprehensible. It was incomplete, which is why it was not until the third period that true quantum mechanics appeared when Max Born, Werner Heisenberg, Louis de Broglie, Erwin Schrödinger, Pascual Jordan and Paul AM Dirac (among others) offered fuller descriptions of a body of physical law which not a single person on Earth could understand.

There was a clear need for some outstandingly impressive mathematics which would still leave the physicists themselves thoroughly bamboozled, thereby keeping up the good work. It arrived in the form of quantum electrodynamics.

In the late 1940s, three physicists, Shin'ichiro Tomonaga, Richard Feynman and Julian Schwinger, working independently of each other, arrived at the same mathematical conclusion at the same time.

It will do you no harm to add that a fourth, the sage-like British-born physicist and mathematician Freeman Dyson, never won the Nobel Prize and should have done so for explaining quantum electrodynamics to the other three.

(There is a great poetic justice in the fact that Tomonaga was based in Tokyo, which narrowly escaped the fate of Hiroshima and Nagasaki by Japan's unconditional surrender in sufficient time to avoid nuclear destruction in August 1945. Just months later, in 1946, Tomonaga sent his workings to none other than J. Robert Oppenheimer, the father of the atom bomb.)

When interactions between light and matter are contemplated, the possibilities are infinite; and it was the 'infinities' in calculations which sent physicists loopy. Quantum electrodynamics (QED), instead of banishing infinity from the sums, invited it to join the party. Miraculously, it solved the problem.

QED is one of the most respected scientific theories ever. As Feynman was fond of pointing out, if you imagine that the calculations deal with the distance from London to New York, they would be accurate to within the breadth of one human hair.

Such accuracy is a good thing to invoke. By 1950 quantum mechanics had become what the physics professor Paul W Davies called '...the most powerful theory known to mankind' and what *New Scientist* termed '...the most successful and wide-ranging theory devised by human ingenuity'.

There were those, however, who felt that the intricate edifice stood on quicksand. Einstein, whose work with light and electrons had opened the curtains on the whole elaborate 'quantomime', wavered between calling quantum mechanics 'incomplete' and declaring its ideas to be 'the system of delusions of an exceedingly intelligent paranoiac, concocted of incoherent elements of thought'.

Obviously, you're in the presence here of a curious blend of derailment and reassurance. There will always be those who feel that confusion is entirely appropriate in God's territory; so you are absolved from all obligations to understand any of it, and need merely pick and choose from such descriptions as you deem worthy of exploitation.

But the incomprehensibility of quantum physics is a secret which you should guard with your life.

MATTER MATTERS

DOES ANY OF IT MATTER?

The greatest potential for paralysing another human being, far better than tickling, getting a submission with a step-over leg lock or administering a generous dose of ketamine, is to introduce a novice to the basic structure of matter.

All matter is made of molecules, you will say, and these can be further divided into atoms. Explain that the atom was accepted as the entity at the basis of all matter from 430BC until 1897, so it is a twentieth-century fashion to talk blithely of 'sub'-atomic particles.

The kernel or core of the atom – the nucleus – is one ten-thousandth the size of the entire atom, and is itself composed of particles (protons and neutrons) bound together by a strong nuclear force. Protons carry a positive electrical charge. Neutrons are electrically neutral.

'Surrounding' this positive-and-neutral nucleus are particles of negative electricity called electrons. (As previously noted, whenever in doubt, bluffers should add an 'on' as a suffix to any word in the quantum vocabulary.)

Try to imagine, you will say again, a 'shell' effect around the nucleus (like the illusion of solidity produced by a whirring top) which is formed by these electrons while zooming at a speed the equivalent of four times round the earth in one second. In fact, of course, they are whirring around a nucleus that is itself only one ten-thousandth the size of an atom.

For a clear idea of scale, thinking of the cricket ball in the cricket ground still applies (and questions of scale represent exceptions to the Don't Visualise rule). You might want to use the image of a half-sized speck of dust inside the Taj Mahal. Others ask people to picture the dome of St Peter's Basilica in Rome. The electron shell would be the dome itself, and within the space of the dome – more or less at the centre – the nucleus would be the size of a grain of salt. Of course, you won't forget to say that the vastness between the nucleus and the electrons is almost completely made up of empty space. If you also remember to add that within the outer shell there may be layers and layers of inner ones, like skins, then the members of the enraptured audience surrounding you will really know their onions. (You can have that particular 'on' suffix for free.)

It is worth stating again that electrons are infinitesimal or wisps, and neither point-like nor planet-like. They flit about like supercharged moths. We are stuck with the term 'spin' and as particles are deflected through magnetic fields their angle of momentum can either be 'up' or 'down' (a typically idiosyncratic physicist's expression for 'one way' or 'another'). It was Wolfgang Pauli who showed that a single particle will insist on being alone and thrumming in its own way in its own energy state. This is called the exclusion principle. If another particle joined it, it would have to have an opposite angle which is called 'spin'. A third would make the others feel exceptionally crowded and would have to go off on its own or find another one with which to have opposite 'spin'. You should limit the ways in which you let anyone say it is just like life.

The balancing of electrical charges at the subatomic level is carried on throughout the universe; all the negative charges balance out all the positive charges. To some, this may suggest that the universe all adds up to nothing. To others, it presents another example of cosmic balance.

You may care to enlarge a bit here by stating that protons are 1,836 times bigger than electrons. Scientists tend to state this precise statistic and then say that neutrons are 'slightly' bigger than protons, without saying how big slightly is. You have little choice but to follow suit, or to forget these particular facts at once.

But this isn't all. Not by a long shot.

Protons and neutrons are themselves composed of still smaller particles. Yes, afraid so. They are the notorious quarks (pronounced to rhyme with 'forks'). They are one of the exceptions to the 'on' rule, and must never be referred to as 'quarkons'.

Protons, neutrons and the quarks within them cluster together to form the nucleus of the atom. It is acceptable to visualise some nuclei as being very bobbly, like little clumps of caviar, or like raspberries. Just a touch more difficult to bite into.

Billions and billions and billions of atoms make up every tangible and visible object, and as we've seen considerable mileage may be had from pointing to something and stating that it is just nuclei, space, and electromagnetism which create the solidity that we see and touch. What seems as still as a breeze block – or building block – is, in fact, deep down inside, in constant motion.

This is perhaps why some especially 'New Age-y' holists speak of everything as being an illusion. (You can refute this with a well-aimed kick.)

Erwin Schrödinger suggested that the entire universe and everything in it – the yaks and yams and stars and suns, the whole caboodle – was created by a single particle on a journey of incredible rapidity and wild complexity. It has to be said that Schrödinger didn't think this for very long. He was known for taking to his bed before and after such pronouncements; he was in a place to which any of us could so easily go.

QUARKS AND THE STANDARD MODEL

Questions about subatomic particles may arise – indeed, you yourself may wish to ask them – along the lines of, 'Just a minute, no one can actually even see these little thingie-whatsits, can they?' Or, 'How do they know?' Or, 'Are drugs required?'

After all, subatomic particles are, well, subatomic: minuscule, point-like entities (variously described as clouds, tendencies, or even dreams) which leave traces among jets of particles when there are collisions in the quantum realm. Some entities may be said to put out energy. Atoms consist of energy and energy is, effectively, invisible. In short, atoms are unlikely ever to be seen; they are detected. No one had actually detected an atom until the 1980s, with the invention of the scanning tunnelling microscope which can magnify what it is looking at 100 million times.

You could proffer the information that down the decades, experimental physicists have used photomultipliers, and have developed voltage doublers, Van de Graaff generators (not the rock band), cyclotrons, Geiger counters, cloud chambers and even massive tunnels miles wide in which particles have been smashed, accelerated, collided, traced, discovered, annihilated, created, presumed or given silly names. But the impression you should be giving is that you live a life of the mind, not of toil at the coalface. You should not deign to spend more than 20 seconds on any of this. (You have a chess game to play against Deep Blue, you will explain.)

For many years the so-called 'standard model' – the orthodox view of what happens at the subatomic level – was composed of quarks and leptons. Be wary of saying 'quarks and electrons' because electrons belong to a wider genre of particle called leptons, so while electrons are always leptons, not all leptons are electrons. Fortunately, no one is ever going to ask you to name the others because this is about as much as any brain can take.

It would have been helpful if all of us could have just left it at that, especially since the nucleus is packed with six different kinds of quark and six varieties of lepton dancing together. That's it, you could have said. The basis of all matter. Excellent. So simple.

Except it is now suspected that quarks are made up of even smaller particles, at present called subquarks, and that the subdivisions go sideways as compulsively as they go downwards. So that there may be not just six quarks and six leptons but a whole menagerie within the Particle Zoo.

Dramatic 'particle smashing' in 1997 revealed the possibility of a strange basic entity, a 'leptoquark', which might simplify matters. To a physicist, a leptoquark is a 'chalk and cheese' or a 'sow's ear silk purse', but it has been ages since anyone expected Nature to do what she's told.

In all fairness, physicists themselves find it difficult to hold on to their sanity in this reductionist's heaven.

In a sad, rather shameful, burst of wayward visualising, quarks were originally given 'flavours' whereby, for the purposes of calculations and the kind of work they seemed to do, they could be told apart.

These flavours are 'up', 'down', 'strange' (originally called 'sideways'), 'charm', 'bottom' and 'top'. (The latter two were originally called 'beauty' and 'truth'.) As you know by now you should thank or blame Murray Gell-Mann for these quirky quark names; the intention was probably to make them sound friendly. When quantum chromodynamics was developed in 1977, the naming thing went overboard. Quarks were given 'colours', which have less to do with tone than flavours had to do with taste. You can get each flavour of quark in red, green or blue, and antiquarks (*see* below, or in another universe) in minus-red, minus-green and minus-blue.

Now, you are increasingly likely to come across books, plays, films or ballets which hitch a ride on the ideas of quantum mechanics. These tend to be vain attempts on the part of artists to appear intelligent. Amidst all of these wrong-headed and often simple-minded efforts, you may find words such as 'charm', 'colour' or 'beauty' liberally deployed and, at times, even used in titles. Those terms were originally nabbed by scientists trying to seem more arty. Soon arty types tried to be more scientific, thoroughly and woefully misunderstanding the concepts. It is your bounden duty to scoff at and stomp on those outpourings which are proliferating wildly and are charmless, colourless, ugly and wrong.

In all fairness, physicists themselves find it difficult to hold on to their sanity in this reductionist's heaven, and have expressed disquiet about the standard model. Its calculations have proved successful, but as Einstein might have said, 'If you were in charge, is that the way you would construct the universe?'

ANTIMATTER

All quantum mechanics has an *Alice in Wonderland* feeling about it. And antimatter is the looking-glass world. Indeed, the existence of antimatter is one of the arguments put forward for the existence of parallel universes. At least, in this universe it is.

Antimatter is matter composed entirely of antiparticles – elementary particles that have the same mass as given particles, but opposite electrical or magnetic properties. Protons are mirrored by antiprotons, neutrons by antineutrons, and electrons – which are negatively charged – by positrons.

The existence of antimatter was first suggested by Dirac in 1930. He was only trying to help. He was formulating a theory for the motion of electrons and produced equations that were beautiful – a major criterion for him – but they described, indeed demanded, the existence of an antiparticle. Such an entity would have the same mass as the electron but with positive instead of negative electric charge. Physicists pooh-poohed Dirac, but Dirac was one of the least pooh-poohable scientists who ever lived. Sure enough, in 1931, the antiparticle of the electron was duly detected. Because of its positive charge it became known as the positron and vast numbers of positrons have been found since.

When the universe formed, equal amounts of matter and antimatter should have formed; but if that were the case, they should have annihilated each other. Now, the observable universe seems to be composed almost entirely of matter. Rather than antimatter. Or a mixture of matter and antimatter. That apparent asymmetry of matter and antimatter in the visible universe is one of the greatest unsolved problems in physics. In short, it is another question for Cern. (They're looking into it.)

When matter and antimatter meet, the mutual annihilation is even worse than the most frightening marriage you know. So if any original antiparticles are still around, they certainly can't be nearby.

Something may have happened, just after their creation, which destroyed them all. It is a treat for physicists when particles and antiparticles collide, and, as they not-so-delicately put it, annihilate to energy. Still more magical is that nanomoment when out of pure energy, particles and antiparticles are created. It's enough to make them feel like Steven Spielberg.

Yet without the help of physicists, this dance of creation and destruction is taking place all the time – in people, phones, walls, toothpaste, rubbish bins and jelly beans. Nothing could exist if it didn't.

THE QUANTUM LEAP

When you speak of the pulsating electrons inside every object, you need to bear in mind that this is what is happening at room temperature. It may be a hectic flurry to you, but to an electron it's rest. Heat them, though, and they really get going. The hotter they get, the more excited they become, and when they reach a certain pitch of excitement and heat they 'radiate'.

With radiation the electron actually jumps or is boosted through different layers around the nucleus inside the atom. This is what Bohr called the 'quantum jump', an expression that gave rise to the 'quantum leap'. You could do some fancy dinner-party footwork with this one. First, what Bohr spotted were jumps down rather than up. (If you really need to know, an electron jumps downwards when radiation is emitted and upwards when radiation is absorbed.)

Second, the term quantum leap has come to mean in everyday language a huge or sudden advance. Quantum leaps do seem to disobey a number of traditional laws of physics; and, relative to the size of the electron, they can involve impressive energy and range. But here is an opportunity for you to earn maximum bluffing points. Strictly speaking, in the scheme of things, a 'quantum leap' is a seriously small event. Indeed, you can happily startle anyone who uses the term by telling them that few events in nature are as minuscule as a quantum leap.

Billions of such leaps are involved in every light bulb, candle, barbecue, soldering iron, combustion engine, cigarette, their respective electrons all receiving and transmitting like one vast telephone exchange – with light, heat, energy and radiation instead of voices on the line.

Great names of the past like Bohr and Heisenberg – as you know – described the behaviour of those minute interacting particles as 'crazy'. Feynman – the physicists' physicist – with his customary American folksiness, called their antics 'nutty'.

It must not be forgotten that scientists are supposed to be – and usually are – very restrained in their use of language. It's just that nothing for them in two-and-a-half millennia has been as shocking.

Light is electromagnetic radiation. It's the kind of fact that gives the mind a lemony taste and then becomes pleasingly forgettable.

SHOCKING BEHAVIOUR

As classical scientists strode out of the nineteenth century, anything scientific was so respected that whatever was unscientific was looked at as if it were a schnitzel that no one had ordered. Indeed, as the twentieth century began, it was believed that all the problems concerning nature and the physical world had been solved by science. Those minor problems that remained would be solved before Wednesday week, at the latest.

Wednesday week came and went. So did many very clever people, and within a quarter of a century all notions of space, time, gravity, energy and matter were changed. It was discovered that things bent, warped, scrunched and dawdled. And this was decades before rock 'n' roll. It was shown that if you went faster than the speed of light you would go back in time, and that the faster you travelled, the more distorted your body would become. Furthermore, the faster you travel, the slower your brain would run. Planes, trains and motorways are full of examples of that.

Perhaps most alarmingly, it was established that if your parents zoomed off at a speed getting on for the speed of light and stayed away for 40 Earth years, they would come back younger than you. And you could get your own back. And so could they.

All these things were accepted by Einstein – after all, he proposed them. But what he could not absorb were the shocks of quantum physics. When asked why, he said: 'A good joke should never be repeated twice.' (Grammar was not Einstein's strong suit. Nor was humour. Nor, as it happens, was maths.)

Within the quantum realm, the rules of nature would be happy merely to be turned upside down. They are turned every which way, and many billions of other ways that are inconceivable to the human mind and are likely to remain so for a long, long time to come. Quantum theory challenges, at the very least, notions of causality, predictability, reversibility, continuity, locality, order, separateness (sometimes called isolability), clear definition, accurate and informative measurement, objectivity, either/or thinking and, most controversially, certainty. It threatens notions of solidity and substance, and suggests that there are absolutely no absolutes.

CAUSALITY

Causality is one of the most important pillars of science. Everyone (or nearly everyone) knows that if you freeze water you will get ice and if you heat it, it will evaporate. But in quantum theory events seem not to have such well-defined causes and effects.

It all boils down to the holist/reductionist argument. Coming down on either side is dicey. Holists are convinced that an electron can jump between orbits for no discernible reason, or that a subatomic particle can come into being or disintegrate altogether without any cause. In other words, holists insist that particles do the loathsomely worst thing anybody or anything can do to a scientist: they behave spontaneously.

Reductionists share Einstein's lifelong distaste for the incorporation into science of anything 'acausal', random or 'stochastic' – a good word to use if you want to say 'subject to guesswork, or conjecture'. He said: 'I cannot believe that God plays dice with the universe.'

If you look at a glass of sparkling mineral water, Coke or, best of all, Champagne, you might begin to see the problem. To analyse or anticipate the behaviour of any individual bubble is fiendishly difficult, even if it is momentarily motionless, let alone when it suddenly decides to jiggle around, disappear or shoot off in funny directions. Quantum physicists have set themselves a task far tougher than relating a bubble to a drink, or even an atom to a bubble.

Reductionists say they have an answer for all that fizz, zoom and kapow!

The real kapow! lies in the fact that their methods deal with particles collectively, via the distribution of probabilities within what they call a macro system, i.e., the whole picture. In other words, their solution is holistic.

Meanwhile, holists say that it's no good calculating collectively: what about the individual particle? In other words, they are thinking like reductionists.

You only need to point to this transposition of views to be protected in any discussion about the stochastic, random or spontaneous nature of subatomic particles.

You will even find a number of physicists who dismiss spontaneity – just like that. For them, the central contentious issue when it comes to causality is time. Is causality linked to time as profoundly as the word 'consequence' is related to 'sequence'? Can an effect precede its cause? It's a question that, of course, arose from the following.

PREDICTABILITY

The remarks about causality also apply to the issue of predictability. Anyone could have seen that coming.

REVERSIBILITY

Reversibility may be understood by watching a video of some movement in nature in what is loosely called real life: a person or an animal running, or swimming, or a speeded-up film of a flower blooming. When such a scene is shown backwards it always looks odd, not lifelike, even funny, because of what is known as the irreversibility of Nature. Natural processes take place in time, and they cannot be 'run backwards'.

However, a video or film of a subatomic particle in motion could run backwards for a century or indeed, conceivably, for several millennia, without looking odd. The viewer might have clues – identifiable events, perhaps, or a peek at the label – but no one coming fresh to that video as it played would be able to identify either the forward or backward motion of it. In short, subatomic particles are somehow outside time – occupying a gap in the very nature that they compose.

Some have been in existence since the universe began. Others have been created as recently as a second ago and might have disintegrated a relatively long time before that second was over. And there is a strong possibility that there are subatomic particles whizzing about here and now that were created in the future.

It is comforting when contemplating these phenomena to realise that even the greatest geniuses are 'up against the limits of what can be known'. This is a handy expression to use on its own or in conjunction with 'up against the barriers of language'.

The latter ploy is better if you want to imply that you do have the knowledge but it is the language that is letting you down. The former is excellent during a highly unlikely moment when you need to be humble. And, almost as unlikely, brief.

CONTINUITY

If you were to arrive in the quantum realm, what you would first notice is the lighting. Strobe, unfortunately. It is an irritating flicker, flashing much more rapidly than in any nightclub. And the particles, like dancers in a nightclub lit by a strobe, are known to be 'here', then to be 'there', but no one knows what happens in between. Their progress is discontinuous.

In classical science it is possible to 'trace a path',

'record a progression', 'monitor a process'; but the lives of subatomic particles are mysterious flickerings, and the bursts of knowing where they are, and what they are doing, are shorter than the gaps of disappearance and unknowability. Like teenagers, really. Not all teenagers are like that, though, and teenagers do tend to grow up. These blighters are arrested at that impossibly awkward stage of not knowing what or why they are, or where they're goin'. Or where they bin.

THE PARTICLE/WAVE PROBLEM

In the quantum universe there is an essential and baffling fact: subatomic particles can also be waves. This you should speak of as the 'wave/particle duality'; you should add that it is the central mystery of quantum physics.

For a reductionist, the mystery concerns the so-called 'quantum amplitude': a particle is of a different size from a wave. It ought not to be one and the same.

To a holist, there is no mystery; it is glorious and natural that nature's tiniest possible entity – a particle – should have connecting ripples throughout the universe.

With particles and waves you see science in a very bad light, and indeed it was in the course of a dispute about light that the major surprises suddenly appeared. Isaac Newton had stated that light was composed of particles. He called them 'corpuscles' and expected everyone to agree with him because he was a genius.

They did agree with Newton until a certain Thomas Young came along.

Young proved conclusively that light was made up of waves. Fortunately for him, Newton had been dead since 1727 and wasn't around to argue the toss. You can seem scholarly by saying that Young conducted his experiment in 1801 or 1805 'depending on the source'.

You can seem even more scholarly if someone suggests that Young deciphered the Rosetta Stone by saying that he did find the sound value of six signs. (And add that Champollion did most of the work and would not share the credit.) Throughout the nineteenth century, everyone was so sure of the wave nature of light that they concentrated on finding out what was waving. As with most romantic plots, electricity and magnetism, which had started out not being suited to each other, were brought together and engaged. Michael Faraday did a lot of the basic matchmaking, then James Clerk Maxwell showed that 'electromagnetism' travelled at the speed of light – and he married them. Light is electromagnetic radiation. It's the kind of fact that gives the mind a lemony taste and then becomes pleasingly forgettable.

While light is always electromagnetic radiation, not all electromagnetic radiation is light. The light that can be seen makes up only a small part of the kinds of electromagnetic rays that there are in the universe, which include X-rays, gamma rays, microwaves and even radio waves.

In this sorry saga so far, Newton had said that light was definitely made up of particles and Young had said that light was definitely made of waves. Maxwell,

Heinrich Hertz and many others agreed: yes, light was made up of waves. That was that, then.

Until, once he'd been working with the photoelectric effect, Einstein stated: 'Nein, light is definitely made up of particles.' He didn't deny light's wave nature, but he could no longer deny light's particle nature either. Which was why the pioneers of quantum physics had to have another look at an old experiment – the one with the two slits, which you should refer to as the double-slit experiment.

Quote Feynman, who said that all discussions about quantum mechanics will eventually come back to the experiment 'with the two holes'. The apparatus consisted of a light, a screen into which two very thin vertical slits were cut, and a wall. In Young's test the light had clearly lapped and undulated its way through and away from the slits, because there were blocks of light and shade on the wall. Particles of light would have flown like bullets, leaving no shading, just two clear slits.

These days, the very same test shows that light is made up of both particles and waves. If you use a particle detector and test for particles you get particles. If you use a wave detector you get waves. Somehow, and no one knows how, it is the test itself that gives you the result. It is utterly alien to classical physics.

Eddington once said that on Mondays, Wednesdays and Fridays an electron is a particle and on Tuesdays and Thursdays it's a wave. (He didn't work at weekends.)

As Walter Cronkite used to say on American TV at the end of the news: 'And that's the way it is this Sunday lunchtime.'

ACCURATE AND INFORMATIVE MEASUREMENT

It is important to make it clear, in a magnanimous and sage-like way, that the measurement of subatomic particles is phenomenally precise. Scientists, especially reductionists, cannot hear that often enough.

The problem is that scientists – like all of us, only more so – need two or more measurements to work together, so that other conclusions may be drawn. If it is known that a plane is travelling at 500 miles an hour and that Rome is 1,000 miles away, then the plane should take two hours to arrive; that can be handy, or essential, to know.

In science it is generally required that data should do more than one job, and in classical science, position and momentum (by which physicists mean velocity times mass, rather than speed) go together like the Blues Brothers, Laurel and Hardy, or trains and annoying people on mobile phones.

The shock when it came to quantum physics was that it was possible to confirm the position of a particle but not its momentum; the particle was too small, too inaccessible and too capable of deflection – as always – by the instrument and the light used to trace it. All concentration was thus focused on finding the momentum. In fact, the momentum was established – with equal accuracy – but by then the particle had zoomed past and its position could not be pinpointed with anything like the same accuracy.

In any discussion about this conundrum you really need do nothing more than utter the three words:

'Heisenberg's uncertainty principle'. It is possible to be certain about one measurement, but then another would become uncertain. By the way, Heisenberg stated that just as it was impossible to ascertain simultaneously the position and the momentum of a particle, so one could not know at the same time a particle's position and its energy. It is rare for this to be pointed out and you will, or should, be highly regarded for doing so.

Inevitably, the uncertainty principle is not principally used by professional physicists for what it has to say about uncertainty. Nor for its remarks about measurements. It is simply a useful tool for some calculations. Scientists do not think that philosophy is a useful tool for anything.

The really terrifying bit of uncertainty for scientists, and indeed for all of Western society, concerns objectivity.

OBJECTIVITY

If you ever get the urge to challenge classical science, state Werner Heisenberg's finding: 'The act of observation affects that which is being observed.' And then watch.

In science, any test must work independently of the person conducting it. The worst put-down one can receive from a scientist is: 'Now you're being merely subjective' – i.e., influenced by individual feelings or opinions, and therefore irrational. Only objectivity, the collective subjective opinions of like-minded scientists, can be trusted. Like a rugby prop forward, objectivity is solid and dependable. And it won't cry.

It is from objective truth that the laws of nature

are derived. And these laws hold good whether or not someone is checking on them. With objectivity, the observer cannot affect the experiment; the experiment and the observation of it are separate in every way.

In the quantum realm it's hard enough to keep the experimental apparatus separate from the experiment itself; there is also the added problem of the physicist always getting the result being tested for. Is the human being profoundly involved?

Believe it or not, you can say that physics has rather moved beyond such questions. You are quite at liberty to point out that the experiment would work just as well while the physicist was having lunch in the canteen.

The physicist wouldn't see it, though, so these questions do linger. Besides, they give you the opportunity to talk about that legendary character, Schrödinger's cat.

SCHRÖDINGER'S CAT

Schrödinger's 'thought experiment' proposed that a cat be placed in a sealed box containing a small flask of cyanide gas which might (or might not) be shattered by a hammer that might (or might not) be activated by an emission from a lump of (potentially) radioactive material – and the experiment be run long enough to give the cat a 50/50 chance of survival.

The cat represents two possibilities, or probabilities. It could be alive or dead and continues to be in that state until – well, until the scientist takes a look. Until then, it is in an absurd and, it has to be said, unscientific state of live-deadness.

OBSERVATION KILLS THE CAT

This story cannot be ignored, but in fact it can be dispensed with very quickly. Apart from anything else, cat lovers never take in the details, they just want to get the cat out of the box and cuddle it.

For some, what is important is that as the act of observation takes place, reality forks and two entire universes are simultaneously brought into being: one in which the cat is alive and scratching, and another in which the cat is dead, deceased and gone to its maker.

Indeed, many boffins go further and declare that the two possibilities represent two separate universes and that, in each of those universes, there may be other twinned possibilities, and other universes. Yes, really – many other universes, billions of other universes.

These are hard-nosed scientists, remember, who are supposed to have a grip on reality. And, by the way, for them, reality – such as it is – doesn't just fork, it fans. This offers up a potentially endless number of realities and is called the 'many-worlds interpretation' – but never say, 'Of what?' The idea of 'many worlds' was dreamed up in 1957 by the American physicist Hugh Everett; you will strike chords with the same notion, in certain circles, by speaking of 'parallel universes'. All's fair – provided you remember that ever-absolving word 'possibly'.

Schrödinger himself went back and forth wondering whether or not he agreed with quantum theory. As noted earlier, he often took to his bed. Equally often, Niels Bohr would follow him into the bedroom and talk at him some more to try to convince him that quantum events were everything and headaches were nothing. It is much regretted that the poor man had such frequent headaches and recurring bouts of illness. In a poetically just universe, he would have been allergic to cats and would have swelled up and started sneezing. Then the rest of us could have observed it – in a vast range of other universes.

LOCALITY

People have been known to scold with the words, 'I can't be in two places at the same time.' Yet it seems as if subatomic particles can. People say it automatically, like, 'How are you?' or, 'This is so civilised', as they pour the tea, or, 'I'm sorry, he's in a meeting'; they just trot it out. 'Of course, subatomic particles can be in two places at once.'

In addition to all the wacky phenomena thrown up by the double-slit experiment, there is the fact – and it is a fact – that a single particle appears to go through both of those slits. It really is as startling as the famously surreal *The New Yorker* cartoon by Charles Addams depicting a skier apparently passing through a pine tree with single ski tracks on either side of the trunk.

The Reverend Dr John Polkinghorne KBE is the kind of person who makes hen's teeth seem plentiful: he's a theoretical physicist who is deeply religious.

Somehow, Polkinghorne's way of putting it is, so to speak, a godsend to bluffers, since he refers to 'that great and good man, Sherlock Holmes', who was fond of saying that when you have eliminated the impossible, whatever remains must have been the case, however improbable it may seem to be. Applying this Holmesian principle must lead us to the conclusion that the indivisible particle went through both slits. In terms of classical intuition this is a nonsense conclusion. In terms of quantum theory, however, it makes perfect sense.

Nothing goes to the heart of the differences between the reductionists and the holists like the question of locality. As far as the reductionists are concerned, classical science has managed more or less to accept and understand that particles of light do become waves of light and vice versa when they have to negotiate, say, the pupils of eyes, or telescope lenses or camera shutters.

And classical scientists never thought to question the fact that such events happened 'locally', at the cricket ground, or at Julia's wedding, or through Paul's window; the light negotiated a small hole, or a thick piece of glass, or a thick piece of glass and a hole, and moved on.

For the holists, there is the knowledge that Heisenberg and Bohr, when they developed quantum theory, began to be aware of something vaguely mystical about it all. Bohr incorporated the Taoist symbol of oneness into his coat of arms. Cult books about quantum physics published in the 1980s created considerable excitement among Buddhists, vegetarians and ageing hippies: 'I mean, don't look for why, man, don't look back, don't explain, know what I'm saying? It just like happens, man. Spontaneously. Why won't my van start?'

This weirdness, or non-locality, is still controversial, and therefore a winner for you. Besides, physicists do their best to shrug off, dismiss or wriggle round any questions that have the faintest whiff of mysticism, so when they take such a subject seriously and address it, so should you.

ENTANGLEMENT

If you want to seem especially knowledgeable and professional when quantum mechanics is discussed these days, you will find few words that are used more frequently than 'entanglement'.

There are some entrancing ways of discussing entanglement; an important thought underlying them is Einstein's conviction that no interactions over any distance could take place faster than the speed of light. Even in his day, though, it did seem as if interactions between particles took place with suspicious simultaneity. Bohr's argument was that separated particles were still part of a single totality, even if one was on Earth and the other on another planet.

To challenge this, Einstein and two younger colleagues, Boris Podolsky and Nathan Rosen, dreamed up a thought experiment, the EPR experiment.

E, P and R believed that they could show quantum mechanics to be 'incomplete'. They suggested an imaginary collision between two particles, which then fly off separately to some distance from each other. Then, by knowing the position of the smash and by probing one of the particles immediately afterwards, all the data about position and momentum concerning both particles could be calculated. Of course, this examination would have to be extremely quick. It would have to happen in a flicker of time. But it would be worth it to contradict Heisenberg and, indeed, to nullify all the shocks wrought by quantum mechanics and save classical science.

There would only be one way in which the EPR experiment could be wrong: if there were some kind of connection or message between particles as if they were two ends of the same stick. Whatever the connection, it would have to 'travel' faster than light (i.e., at a speed that was 'superluminal'). In that event, it would not be possible to probe one particle without simultaneously probing the other. Such a notion was worse than disturbing to Einstein; it made scientists look like fairground psychics, gamblers or dabblers in voodoo. He called it 'spooky and absurd'.

Einstein died. Bohr died. Then, in 1964, John Bell at Cern in Switzerland published some challenging calculations, which you should call Bell's Theorem. Bell had actually begun work in the hope of proving Einstein and his young assistants right, but his theorem opened the door to Einstein's worst nightmare. Repeated experiments, especially during the late 1990s, confirmed that once two particles have had any interaction, they do somehow remain linked as parts of the same indivisible system. Researchers seem perpetually on the brink of putting Bell's Theorem beyond question – especially, in this case, at institutions away from Cern. Where they're also looking into it.

You can expect discussions about the connection between twins, the collective unconscious, TV ratings – any number of subjects can be swept into this gigantic dustpan.

If entanglement doesn't bring out the storyteller in you, nothing will. You'll find your own way of telling this, but your tale requires anything that comes in pairs, one left and one right (therefore not earrings), such as shoes. Einstein used gloves. The Timothy McSweeney website explored the glove analogy too. As with all tales, you will adapt it to suit your style. Be sure to make it clear that your story is describing occurrences that are 'a bit like' phenomena in nature.

You are telling a magical tale. You put a right-hand glove in one box and a left-hand glove in the other. Wrap them if you like. Then, go on a journey, taking one of the boxes, so that the gloves are far away from each other; go even to the edge of the universe.

Then open your box. Inside it you will find a right- or left-hand glove. At that moment you will be able to see whether it is a right- or left-hand glove. And you'll instantly know from that information whether the other glove, in the other box, some distance away, is a left- or right-hand glove.

The two gloves in this story are said, by physicists, to be 'entangled'. One can know certain things from the one by looking at the other.

But now you can notch up the enchantment. Involve a lover. Imagine that you and your lover wrap and box separately a left-hand glove that is bright pink and a right-hand glove that is dark green.

You and your lover travel far from each other, even to opposite boundaries of the universe. When you reach your destination, you open your box and unwrap your glove. A trillion, trillion, trillion gloves swirl around you: left and right, different shapes and sizes, different colours. You reach up into the cloud and pick one glove. It is dark green and it fits your right hand perfectly.

And, because the clouds of gloves are entangled, because they're connected in some spooky but nonetheless real way, you know for certain that on the other side of the universe, at the very moment you're flexing your fingers in your glove, admiring the back and front of your hand, a bright-pink glove has just been slipped onto your lover's left hand. That really is how entanglement is said, by some, to operate in the quantum realm.

Many people love Einstein as if he were their puppy.

Now, however smug it feels to have proved Einstein wrong and to show how spooky the cosmos really can be, you should not allow this to show. Firstly, it would be unseemly. Secondly, you would be hated for it; many people love Einstein as if he were their puppy. Thirdly and most importantly, the questions that Einstein posed and the doubts that he expressed have not completely gone away. You should never stop stating, gently, that quantum mechanics is incomplete. A situation that, needless to say, is being addressed whenever the machinery consents to work at Cern.

Some researchers claim to have shown that the

separated particles 'made the decision to interact in the past'. This saves entanglement from being mystical or spooky. It's called 'retrocausality'. In short, time travel...

ORDER

It is worth noting, for conversations in which others do not mind getting technical, that all electrons are identical to each other, all protons are identical to each other, and all neutrons are identical to each other. In other words, three basic entities, by combining to form atoms, which combine again to form molecules, combine and combine again to build everything in the universe from bratwurst to battleships, from waffles to warthogs, from neck braces to New York. But it will also by now be clear that in every cubic millimetre of space there are billions and billions of subatomic particles which are capable of behaving at times in astonishing and unpredictable ways for no rhyme or reason. These repeatedly, spontaneously and 'nuttily' change their physical qualities, their whereabouts, their energies and their flight paths and, small as they are, seem to have the entire universe in tow. It is quite difficult to put them in neat piles, label them and file them away.

CLEAR DEFINITION

The point never stops being made, unfortunately, that particles that can clearly be defined as particles can also be found to have none of the qualities that are recognised as particle qualities; at such times they have all the qualities of waves and must be defined as waves. This is not a new idea, and yet the shock waves from it still do not seem to have subsided.

It challenges classical science, because clarity of definition is a cornerstone of science. It is almost the definition of it.

SEPARATENESS

Particles are often defined as 'discrete' (ibid.). This is a good word to use. 'Discrete' means 'individually distinct', 'separate'. Holists believe that each particle tugs at and is tugged by the universe. Reductionists cannot quite budge from their stand of separateness and splendid isolation.

Whether you support one or the other (or neither), you need to concede that it is possible to study and conduct experiments with individual particles. Indeed, the ability to send a single electron whizzing through silicon is the very basis of the entire computer industry.

But, when trying to study subatomic particles, it is impossible to subject them to traditional, classical scientific examination, because particles relate to each other and to equipment itself with such determination. It is sometimes stated that these particles are not experimentally 'isolable'. Neither is it possible to separate off any of the shocks that quantum theory has given to physics, to classical science and to civilisation – challenges that arrive from every direction, at all times, simultaneously.

It might seem obvious that this should be so, but Einstein didn't like it. Einstein didn't like it at all.

TITANS AT BREAKFAST

In 1927, during the Solvay Conference in Brussels, held in the Metropole Hotel, Bohr and Einstein conducted a debate which you might choose to call the greatest intellectual encounter since Newton's famous clash with Gottfried Leibniz. Using the word 'famous' when people are unlikely to have heard of something is an unkind but delicious ploy. It helps the story along if you mention that they conducted their arguments outside the business of the conference, and always at breakfast time.

Einstein was in effect defending the Newtonian universe against all those shocks with which you are now familiar.

Bohr narrowly won the first encounter, in 1927, by gently insisting that subatomic particles cannot avoid being affected by the very equipment with which they are examined. It might go against Einstein's deepest convictions, but obviously there are particles in the instruments. Surely they must affect each other.

The second bout took place in 1930, during that year's Solvay Conference, in the same breakfast room of the same hotel.

Einstein had made up a 'thought experiment' which concentrated, subtly, on a tiny particle's tiny energy. As they finished their croissants, it looked as if Bohr had lost.

But at yet another breakfast, Bohr proved Einstein incorrect. Bohr was able to show that the equipment would have to be affected and this time did so triumphantly and dramatically, referring to gravitational fields. Put simply, Bohr defeated Einstein by using Einstein's own theory of relativity.

Kind scientists might say: that Bohr was convincing the great man that quantum theory is not inconsistent with relativity. Bluffers may want to accept that benevolent view. Harder hearts may wish to push the point that Einstein was wrong and lost by scoring an own goal.

Niels Bohr and Einstein kept up their argument for three and a half decades. Even after Einstein died, Bohr would take issue with him as if he were still there, as a warm-up. The last thing Bohr did before his death in 1962 was to draw a diagram of that thought experiment proposed 35 years earlier in Belgium by Einstein, questioning and reassuring himself one last time.

EITHER/OR THINKING

All the new shocks and questions have not shaken the fundamental difference between holists and reductionists.

Reductionists say: 'either/or'.

Holists say: 'both/and'.

You say: Both can be either right or wrong or, indeed, both right *and* wrong.

This applies to both 'either/or' and/or 'both/and'.

CERTAINTY

Obviously, if definitions, accuracy and predictability are at risk, certainty will be in trouble, too. Note:

1. The uncertainty principle refers to the phenomenon of uncertainty with regard to the position and momentum of a particle. It is not saying that nothing is to be trusted anywhere anymore.

2. Classical science is still, for the most part, accurate and certain. You need to say this, even if you do feel moved to add that reductionists need to accept some doubt, some spontaneity, some life.

All possible or probable jokes about the uncertainty principle have been made and made and made again. You can be sure of that.

According to the dreadful old joke, W and Z particles beg the Higgs boson to come to the cathedral on Sunday because without it they can't have Mass.

GUT INSTINCTS

In the quantum universe the end of physics will be found. And thereby, the solution to everything.

While it is clear that there should be a 'grand unifying theory' (GUT), and there have been any number of attempts to prove that there is, all such comprehensive proposals remain satisfyingly loose.

There is no reason why you should not say that you have joined the legions of super-ambitious physicists in the search for this 'theory of everything' (TOE) which Einstein dreamed of finding. (In Einstein's day, a GUT or TOE would have been called an UFT, for 'unified field theory', because it had been discovered that four kinds of 'fields' are permeating the physical universe, and that, sportingly, they allow four 'forces' to frolic there.)

Quite late in their lives Heisenberg said that he and his friend Wolfgang Pauli had just about 'got there'. Pauli, however, drew a cartoon of an artist, brush in hand in front of a canvas on which there were a few lines and squiggles. The caption read: 'I can paint like Titian; only a few details are missing.'

This should inspire you to prepare and cook your own GUT. You are ready to do so.

First, catch your particles of matter. These can be difficult to clean and separate, so just cram them in the pot and persuade as many as you can to stay.

If you thought particles of matter were unwieldy, wait until you try dealing with messenger particles.

Then get hold of your messenger particles. Be warned: if you thought particles of matter were unwieldy, wait until you try dealing with these.

You will find them in the forces of nature which have been marinating at least since the Big Bang. In ascending order of their powers of attraction or repulsion the forces are:

1. gravity;
2. the weak nuclear force;
3. electromagnetism; (remember how electromagnetism, even though it was united with the weak force, is still strong enough to repel shoes on streets and seats on chairs and hands on tables).
4. the strong nuclear force, which you should always refer to as 'the strong force' – the immense tug at the core of atoms. Nothing in the universe binds like it. Its particles are called gluons. (It is the breaking

of this binding that makes nuclear weapons wreak havoc, and the harnessing of it that creates nuclear energy.)

By the time you have stewed, minced, mashed, stirred, whisked and added a pinch of salt, you would hope that they would all be reduced – after straining – to a single force. But be prepared for disappointment. A GUT has never been achieved before (nor has it been grand or unified) for the simple but persistent reason that gravity has been left out and even the other three have not yet been united.

DIPPING A TOE

The major stumbling block is that subatomic particles have always seemed free of gravity, flying about with as much abandon as international businessmen. More to the point, the maths makes utter nonsense.

You can show how difficult this problem is by saying that Einstein laboured on it for decades, trying to unite gravity with the other three forces, so that quantum physics could coexist with his general theory of relativity, and subatomic particles would make sense within the universe as a whole.

Uniting the weak nuclear force with electromagnetism was only achieved some 20 years after his death. Since electromagnetism has a vastly longer range, operates more symmetrically and is 10 billion times stronger than the weak nuclear force, you should regard the discovery that they are the same fundamental force as spectacular, and call the theory that proves it the 'electroweak theory'. Particles that carry this combined force are called, inventively, W, Z+ and Z-, which excites reductionists and no one else.

Two of the men who came up with this were the physicists Steven Weinberg and Sheldon Glashow, who had been friends since their days at the Bronx High School of Science. Still, they worked more or less independently – which excites holists. Also labouring apart but simultaneously was Abdus Salam, who remains without honour in his own country, Pakistan, because his family belonged to the wrong community.

(As a matter of interest, two people from Pakistan have won Nobel Prizes: Abdus Salam and Malala Yousafzai.)

So much for what has been done so far. For the rest, the belief is that the more billions that are spent on them, the more answers they will find. A superconducting supercollider in the Texan desert, for instance, cost the Americans mega-dollars (think billions, then think double that) before Congress pulled the plug when 14 of the intended 54 miles had been built. Nobody mentions it any more. It would have been nearly three times as powerful as the Large Hadron Collider at Cern. (Where, of course, they're still looking into this.)

Nor can discoveries always be as newsworthy as the Higgs boson.

(By the way, bluffers might be interested to know that Peter Higgs may live in Edinburgh but he is not a Scot. He was born in Tyneside and most of his early years were spent in Bristol, where he attended Dirac's old school, Cotham Grammar.)

Electroweak particles (W and Z to their friends) are puzzlingly dense. According to the dreadful old joke, W and Z particles beg the Higgs boson to come to the cathedral on Sunday because without it they can't have Mass. And indeed, with impressive creativity, those entities are somehow, perhaps miraculously, endowed with mass, which is why the Higgs boson was called 'the God particle' for a short while.

On 4 July 2012, experimental physicists at Cern announced that they had found a particle which very jolly nearly was, at the very least, Higgs-like. It was then that the world came to know of the expression 'Sigma-5'. Sigma is a Greek letter. Sigma 1 to 4 do represent the close-but-no-cigar results, but this is the world of science, remember, so 3.6 sigma means 99.9% certain; 5 sigma means 99.99995% certain. There is a distinct possibility, however, that it will be adapted to mean 'pretty sure'. In answer to questions about used cars, damp courses and contraception, stand by for the firm, less than trustworthy response, 'Sigma-5'.

In 1993, William Waldegrave, then the UK government's science minister, awarded bottles of Champagne to physicists who could explain the Higgs boson and why money should be spent looking for it. The winning answer was in the form of a metaphor involving Margaret Thatcher, the recently departed prime minister, making her way through a cocktail party. As political hangers-on cluster around she finds it harder to move across the room – acquiring mass due to the 'field' of fans, with each fan acting like a single Higgs boson.

Some scientists have tried to put across the image of

particles as flour being bound together, but they appear to have confused the name 'Higgs' for the word 'eggs'. Another overworked explanation involves a tray of sugar or treacle and several ping-pong balls. Such a tray would represent what is known as a Higgs field and the particles within that field, Higgs bosons, give mass to the balls as they make their way along. You can trump such explanations by stating loftily: 'Higgs himself loathes such metaphors.'

You can add that he especially dislikes the idea that a Higgs field slows things down. However treacly the environment may be, particles power through it. Peter Higgs is in his eighties, but still rather youthful, and he's possibly defensive about pace of movement.

If you're feeling less than charitable, you can add that Higgs was not by any means the only – nor even the first – person to conceive of these mass-giving phenomena. You can trump the media by calling this entity the 'Englert-Brout-Higgs-Guralnik-Hagen-Kibble' field.

It mustn't be forgotten that the Higgs boson is even more magical than the acorns from which oak trees grow. It is a messenger particle and yet it gives mass to other particles. For what it's worth, Higgs himself sees the Higgs field and the Higgs boson as refracting – shining, as it were. Don't forget that light is made up of bosons. And so the Higgs boson might be described as giving off a blue light that colours everything blue; then, just as plants use light to grow, tiny particles gain mass from that light.

Another tactic is to duck or dodge the details and simply say that the Higgs boson gives mass to particles and thereby matter to everything from people to planets.

WHAT CERN IS ADDRESSING

A few questions (more than a few questions) are being looked into, you can say with quiet confidence. The following are some examples:

- What's with the Higgs boson?
- What's with matter and antimatter?
- Is the standard model substandard or hunky-dory?
- Do all known particles have supersymmetric partners? Or are they only good for one-night stands?
- Are there extra dimensions? And are Grandad's glasses there?

- Are there additional sources of quark-flavour mixing? If so, why won't Waitrose stock them?

- Are electromagnetism, the weak nuclear force and the strong nuclear force different manifestations of one universal, unified force? And is it with us?

- Why is the fourth fundamental force, gravity, so many orders of magnitude weaker than the other three fundamental forces? Can we finally be cleverer than Einstein?

- What are superstrings and are they in that drawer?

- How modest should we be about our ignorance of so much of the universe? Not humble at all? Exactly.

One of the most controversial projects at Cern involved the neutrino. Enrico Fermi suggested this word (Italian for 'little neutral one') for particles that have no electrical charge, are subject to the weak nuclear force and are capable of passing through a block of lead which stretches from here to Jupiter. What role they may play in the picture of the universe as a whole is still up for grabs – or not, as the case may be. Pauli offered a bottle of Champagne for anyone who could prove that they existed. He paid up.

During 2011, experimentalists at Cern announced repeatedly that they had created many bunches of neutrinos and had sent them through 730km (454 miles) of rock to a giant detector at the INFN-Gran Sasso laboratory in Italy. They claimed that 15,000 separate measurements, spread out over three years, showed that the neutrinos arrived 60 billionths of a second faster than light would have done, travelling unimpeded over the same distance.

Einstein had insisted that nothing can travel faster than the speed of light.

Perhaps the neutrinos, like so many before them, had ignored the universe's absolute speed limit once they'd crossed the border into Italy.

Then in March 2012, in a repeat experiment, the neutrinos went fast, but not that fast. The timepieces were checked and found faulty, thus rather embarrassing the Swiss who are known to pride themselves on efficient time-keeping, although it was an Italian who led the research and who, almost at the speed of light, was asked to resign. Generally speaking, though, Cern is thriving, firm in the conviction that the day's not far off when the world will finally have an all-encompassing theory of everything and an equation that can be printed on a T-shirt.

The novelist Martin Amis has suggested that such revelations lie 'five Einsteins away'. It's not a bad trick to come from a position of relative ignorance and spout smart-sounding stuff like that.

You are well within your rights to be somewhat sniffy about superstrings and M-theory, in the interests of science – and in a bitchily even-handed way.

SUPERSTRINGING US ALONG

Supporters of superstrings and M-theory believe that the ultimate truths are coming closer, courtesy of Cern, and that they have one or two Einsteins on the case. String theory and M-theory have been known to bring physicists the closest some will ever get to visible manifestations of emotion. As with all GUTs, TOEs and UFTs, they will answer questions about the beginning of time, the origin of the universe, whether there are multidimensional universes and the best way to make a cup of tea.

The idea is that the fruits of centuries of enthusiastic investigations – the totality of information about the universe – will be encapsulated in a single equation; and that if the right equipment was available, it could be seen that the ultimate fundamental entities are minuscule filaments, vibrating strings or small membranes. (When you say 'small', know that those strings look up at an atom as an atom looks up at the entire solar system.)

Superstrings are believed to thrum like the strings of a musical instrument. Instead of notes, the strings are supposed to produce streams of particles, and thus all of creation is somehow tuned into existence.

Michael Green and John H Schwarz had worked out string theory by the late 1960s and over the next decade and a half, they managed to sort out such mathematical inconsistencies as were obvious at the time. In 1984, before the days of email, they FedExed a letter about their precious strings to Edward Witten, a brilliant young physicist and an exceptional mathematician at the prestigious Institute for Advanced Study in Princeton. Witten was smitten. He was already an influential fellow and wires all around began to buzz about strings; they have buzzed unceasingly since.

For Witten and cohorts of other researchers, calculations have a better chance of attaching to vibrating thread-like entities than to minuscule, not properly described 'point-like' particles, which send their numbers into the dreaded world of infinities. Also, it is a great scientific insight that superstrings carry not just the strong force, but all forces. So superstring theory doesn't just propose that gravity is included. It requires it.

M-theory stands for mother of all strings, magic, mystery or membrane theory – though some of late have called it madness. It proposes that strings exist in various dimensions, as membranes or webs. The membranes whose dimensions exist as no dimension are called 'zero-brane'; the ones that have string-like dimensions are called 'one-brane'; those with dimensions like a bubble are called 'two-brane' and so forth. At one stage it was half-seriously suggested that the word 'p' for particle might establish a category called p-brane.

Imagine the kudos you can earn when talking about extra dimensions. People will believe you can actually perceive them. Visualise them if you like... You can't. They exist in the sums and, the theory goes, they exist (in effect) non-existently in real life.

Hypothetical, multidimensional regions are named Hilbert spaces after a great mathematician, David Hilbert. Anyone mentioning Hilbert spaces sounds impressively like a professional physicist; and to appear even more knowledgeable, it is worth lamenting the fact that they are not named after the mathematician John von Neumann (pronounced Noymun).

Von Neumann's master stroke in 1929 was to adapt Hilbert's ideas, providing physicists with spaces that have an infinity of dimensions. Pictures of Hilbert spaces tend to look like the tangle of proliferating pipes outside the Pompidou Centre in Paris – happily without the white-faced mimes and also without someone selling T-shirts saying 'My physicist friend went to a Hilbert space and all I got was a T-shirt saying, "My physicist friend went to a Hilbert Space and all I got was a T-shirt saying 'My Professor went to a Hilbert Space...etc.'

You are well within your rights to be somewhat sniffy about superstrings and M-theory, in the interest of science – and in a bitchily even-handed way.

You should be able to comment on a book by the American physicist Lee Smolin entitled *The Trouble with Physics*. He may as well have called it *The Trouble with*

Physicists, but you definitely shouldn't go there.

It used to be said that testing string theory would require 150 years and a microscope the size of a galaxy. Then things seemed to be getting worse. Some researchers declared that string theory cannot be tested at all, nor proved, nor falsified. Murray Gell-Mann appeared to be holding out, asserting that the elements of string theory most certainly be tested, so that the whole theory can be confirmed or denied experimentally. Still a growing number of brave scientists say it is hurting science. Too many hypotheses, too much hype.

Initially, string theory was mathematically elegant, even beautiful. Now, after decades of manipulating and mangling the figures, neither the numbers nor the universes they are seeking to represent are attractive, not even to motherly types.

Critics of string theory argue forcefully that its proponents have ladled into their bowl an indigestible gruel containing an extra six or seven or 26 dimensions which no one could ever see, trillions and trillions of other parallel universes which no one could ever visit, tunnels through space and time, and other creepy-crawly ideas taking string theory from being a theory of everything to being a theory, quite possibly, of nothing.

You're quite at liberty to use the expressions 'superstring theory' and 'string theory' interchangeably and you can certainly use the plural form. Without necessarily dismissing string theory entirely – a dangerous ploy – you might gently point out that, because of all those hypothetical universes and far-flung notions, string theories have proliferated to such an extent that there are now effectively more string theories than there are atoms in the unknown universe. Where calculations pertaining to string theory can be judged – and there are very few of them for science's liking – one prediction made by string theorists is so inaccurate that it would be like saying that a baseball is the size of the sun.

Tough as it may seem, you should be even-handed here and point out that at such a scale, dramatic deviations in the mathematics always occur. String theory is an entirely new science and by definition it breaks all the rules; the connections among those hypothetical strings are different, deep down, and so distant from the world around us that completely different approaches are required.

One prediction made by string theorists is so inaccurate that it would be like saying that a baseball is the size of the sun.

It is worth betting on more breakthroughs. It has had some astonishing successes and so an entirely new approach is required to understand it. There will be some spectacular surprises and if they could be described now, they wouldn't be surprises, would they?

Never forget that a number of geniuses support string theory. Gell-Mann called it the only game in town – some time ago, admittedly. Still, Green, Schwarz and

Witten are not slouches and should not be opposed lightly. Top-class scientists tend not to work on failed ideas; no guns are pointed at them and there is no barbed wire between fields of physics. Real, true, heroic science involves discovery, wonder and being prepared to accept what seems unimaginable. It's not that string theorists love the scientific method last; they just love science more.

If you're brave enough to do so, you could hold fast to the many strong objections to string theory. Physics departments that look to other theories don't get funding; and yet string theory might have shattered the hopes and dreams of thousands of brilliant scientists and dedicated people in those departments. And there are powerful witnesses against these ideas, including no less a figure than Richard Feynman, who remained disapproving and unconvinced. Also unimpressed is the renowned Nobel Prize-winner Sheldon Glashow, who penned some doggerel verse, enjoining us all to heed good advice, lest we all become smitten: 'The book is not finished, the last word is not Witten.'

SOKAL'S HOAX

There is a great deal of satisfaction to be had from this story because it goes to the heart of the Two Cultures discussion. You can maintain your superiority throughout your telling of it, giving yourself great boffin-like authority, if you show how ridiculous humanities numpties can be.

In 1994 a New York physics professor, Alan D Sokal, sent a paper to a hitherto respected cultural studies journal, *Social Text.* His piece, which was published in 1996, was deliberately full of scientific howlers. None of the academics who read the article had sufficient knowledge to correct his work and it was published. He showed that even lefties like himself would accept a paper that agreed with their own politics and would make it public, if it had the imprimatur of science.

Sokal made up a rambling title: 'Transgressing the Boundaries: Towards a Transformative Hermeneutics of Quantum Gravity'. The piece itself was chock-full of statements that went way beyond the merely bonkers. Sokal talked of the 'external world, whose properties are independent of any individual human being and indeed of humanity as a whole'. Even first-year physics students would roar with laughter at his claim that 'the [Pi] of Euclid and the G of Newton, formerly thought to be constant and universal, are now perceived in their ineluctable historicity'. He pushed the gobbledygook on and on into Drivelsville: '...and the putative observer becomes fatally de-centered, disconnected from any epistemic link to a space-time point that can no longer be defined by geometry alone'.

Sokal followed up his paper with a book called *Fashionable Nonsense*, written with the Belgian theoretical physicist Jean Bricmont. Here, the targets were authors in France, a veritable pantheon of contemporary French 'thinkers' whom you can safely and comfortably debunk. These swaggering figures include: Jean Baudrillard, Gilles Deleuze, Félix Guattari, Luce Irigaray, Julia Kristeva, Jacques Lacan, Bruno Latour and Jean-François Lyotard.

You should concede that they couldn't pin anything on the philosopher Jacques Derrida, who wrote a critique called 'Sokal and Bricmont Aren't Serious', first published in the French newspaper *Le Monde*. He commented in passing that the Sokal hoax was rather 'sad [grave]', not only because Alan Sokal's name would now be linked primarily to a hoax, not to science, but also because the chance to reflect seriously on the issue had been lost to a broad public forum that deserved better.

Notwithstanding the lack of seriousness, each and every one of the overblown figures identified by Sokal and Bricmont were condemned in the most embarrassing way possible, by simply quoting verbatim what they themselves had written. They had tried to make themselves seem clever and to bolster any number of wayward world views and prejudices, but all those kings and queens were all together in the altogether.

Julia Kristeva really did write that 'poetic language can be theorised in terms of the cardinality of the continuum', Baudrillard did assert that 'modern war takes place in a non-Euclidean space', and Latour honestly did muse about what Einstein had learned from him.

Lacan really did state that the erect male member is equal to the square root of minus one.

These words really were written by notionally sentient beings: 'In the Euclidean space of history, the shortest path between two points is the straight line, the line of Progress and Democracy', as well as 'Is $E=mc^2$ a sexed equation?' Lacan really did state that the erect male member is equal to the square root of minus one.

Clearly, like particles gaining mass from the Higgs boson, those who cannot measure up intellectually to physicists are trying to hoover some scientific credibility into themselves, while at the same time disparaging science for being too much of a bully. The suggestion is that science is an oppressive force. But truth does not oppress anyone. Science at its best liberates people. It is lies that oppress us.

You have a golden opportunity to ensure that science does not win hands down because of this scandal. First of all, in order to get his article published, Alan Sokal was quite happy to exploit his own reputation and authority, so his experiment was not 100% pure.

Secondly, thus far no one has dared to write such a thing, but it's a fair bet that a carefully composed and suitably zany theory in a certain scientific field would get peer reviewed and published.

The time is ripe for someone to write a barking-mad paper, as crazy as Sokal's, involving string theory.

GRAVITATIONAL WAVES

Einstein first predicted the existence of gravitational waves in 1916. Remember that Einstein's Theory of General Relativity deals with space and time and everything gravitational.

It is important to understand that gravity waves are not gravitational waves. Both have gravity in common, but they are different. Examples of gravity waves include wind-driven waves in water, as well as tides and tsunamis. Gravity waves are physical perturbations driven by the restoring force of gravity.

Gravitational waves can bring out your storytelling gifts again. *'A long time ago in a galaxy far, far away, two massive black holes engaged in a dark dance of death. Spiralling faster and faster, reaching half the velocity of light, they collided. During that cataclysmic encounter, they merged to form a doubly massive black hole. A tremendous event. It reverberated through immense distances. It continued to do so for aeons.'*

In other words there was a gigantic crash that had an impact upon space and upon time – which you should call space-time. And such distortions of space-time, such ripples caused by violent and energetic processes in the universe, are what you should call gravitational waves.

Almost a century after Einstein's prediction, on 14 September 2015, the universe's gravitational waves were observed by human beings for the very first time, by perhaps the most sensitive instrument ever devised on Earth, the Laser Interferometer Gravitational-Wave Observatory, which you should call LIGO and which is a very useful thing for bluffers to know about – not least because its acronym should more accurately read LIGWO. (Physicists can be appalling pedants too.) LIGO's teams work in the US from two main sites: one in Livingston, Louisiana and the other in Hanford, Washington, 3,038 kilometres – 1,890 miles – apart. LIGO cannot function alone. It has to have an equally sensitive partner. In common parlance it has to 'reach out'.

Bluffers are quite entitled to make exorbitant and exaggerated claims about LIGO, declaring the destruction of umpteen dark energy theories, affirming ideas about gamma rays as well as basic elements and even perhaps offering a GUT (see page 71) to the effect that as the universe began all four forces were united and only split up later because of irreconcilable differences. That's what LIGO has done for us.

DON'T SAY: Where's the telescope?
DO SAY: LIGO is 'blind'. LIGO can't see anything and doesn't need to. LIGO doesn't use telescopes. Rather, it listens out, seeks to detect invisible gravitational waves. It 'feels' for them.

DON'T SAY: Why's it not round?
DO SAY: See what you can remember of the following: Each LIGO site uses two straight, level vacuum tubes, made of steel, 1.2 metres – about 4 feet – in diameter. They stretch for 4 kilometres – two-and-a-half miles, making it the seventh-longest building in the world. They're protected by a 10-foot wide, 12-foot tall concrete casing and are arranged in the shape of an 'L'.

DON'T SAY: As in L for Ligo?
DO (BE TEMPTED TO) SAY: They do kind of work like ears, giving their full attention to the cosmos.

The ripples from the collision of those black holes took 13.7 billion years to reach LIGO. It took a hundred years for Einstein's conjecture to be confirmed.

And the quest took up the lives of hundreds of dedicated researchers, all of whom were acknowledged when the leading lights of LIGO won their Nobel Prizes in 2017. Half of the $1.1 million award went to Rainer Weiss and the other other half was split by Barry Barish and Kip Thorne.

Rainer Weiss (1932 –), Rai to his friends, of whom you're clearly one, has been dreaming of detecting gravitational waves for half a century. He works hard. His swift response to winning the Nobel was a thoughtful: 'It will f*ck me up for a year.'

Interesting bluffing fact: Weiss went to the Columbia Grammar and Preparatory School in New York. At the same time there was a more senior boy at the school,

Murray Gell-Mann, with whom the school's teachers compared Weiss unfavourably.

(Equally interesting, Dirac and Higgs – as you can see elsewhere in this book – also went to the same school, though in their case not at the same time.)

As it happens Weiss had been something of a drop-out, a fool for the love of women and an habitué of workers' bars, but he had tremendous practical skills. In 1976 he was elected chair of the group working with NASA's Cosmic Background Explorer or COBE, which began orbiting the Earth in 1989 and quickly produced our first insights about the new-born universe. Weiss was overlooked for the Nobel Prize for that so 2006 wasn't a f*cked-up year for him.

LIGO LAUREATES

Rainer Weiss is based at the Massachusetts Institute of Technology, MIT. Barry Barish and Kip Thorne are at the Californian Institute of Technology, Caltech.

Barry Barish has also had connections with the Nobel Prize. In the early 1970s he co-ordinated a project at Fermilab involving the colliding of neutrinos with the intention of finding out more about quarks. But what was also revealed was a weak neutral current, which became the linchpin of the electroweak theories of Salam, Weinberg and Glashow.

Barish has said that he used to long to write a great novel but science – and presumably cheerfulness and sense – got in the way. Probably just as well. The third laureate in this group, Kip Thorne, has said, 'Barish, in my opinion, is the most brilliant leader of large science projects that physics has ever seen.'

If you can judge a man by the company he keeps, then Thorne's long-time friendships with Carl Sagan and later Stephen Hawking should tell you a lot.

In 1975, Weiss invited Thorne to speak at a NASA committee meeting in Washington, DC. Hotel rooms that summer were fully booked, so the two shared a room and stayed up all night talking gravitational wave theory.

Thorne came away and proposed creating a group at Caltech to work on the detection of gravitational waves using interferometers.

They have grown old, all these men, now in their 70s and 80s. Their universities call them Emeritus, which effectively means retired (but still with it). A crucial fourth member of the team and co-founder of the LIGO project, Ronald Drever, died between the triumphant detection of the waves in 2015 and the announcement of the Nobel Prizes in 2017.

Nobels are not given posthumously, but if Drever had lived longer he and not Barish would probably have won the Nobel.

Award yourself maximum bluffing points for knowing the following: Kip Thorne is the person who gave the world the idea of the wormhole, the theoretical passage through space-time. And it is Kip Thorne's story, and his ideas and influence that are behind the movie *Interstellar*.

The Italian physicist Enrico Fermi said that if he could remember all the names given to particles he would have been a botanist, and if such things really do interest you, you need to get out more.

FAMOUS PHYSICISTS

Whenever the conversation gets a bit sticky, it is well worth your while to have some snappy facts about the People Who Made the New Physics. This is especially helpful if no one else has heard of them and you can blind everyone with scientists.

JOHN BARDEEN (1908–1991)

John Bardeen is the only person in history to have received two Nobel Prizes in physics. Bardeen should be far more famous than any number of loud-mouthed scientific hype-meisters, but 'Whispering John' simply did his work and never bothered with boasting. At Bell Labs in New Jersey, the golf-crazy Bardeen and Walter Brattain, with contributions from William Shockley, succeeded in creating the first transistor.

Maximum Bluffing Value (MBV) In 1956, just before his first Nobel ceremony for that work, Bardeen had to borrow a shirt, his own being deemed unsuitable for a major occasion. To make things worse, when King Gustav of Sweden presented the award, he scolded Whispering John for leaving his children behind on such an important occasion.

In 1972, Bardeen won his second Nobel Prize for being in the group that pioneered work in the field of superconductivity. That time he brought the kids.

DAVID BOHM (1917–1992)

Recording chats with the Dalai Lama and Krishnamurti, Bohm was one of the holiest of the holists, and you need to be aware of him. He was sure that there is an underlying unity to all physical, psychological and spiritual experiences. This unity is as yet indivisible; he spoke often of 'hidden variables' – a notion that is gaining popularity among physicists.

For him, waves and particles were not always evident but were always present, the waves somehow guiding the particles. Bohm's calculations in the early 1950s led to later proofs of subatomic non-local quality, but then he lived a non-local life in the USA, the UK, South America and the Middle East.

MBV Bohm spent long periods of time living among the Blackfoot people in what reductionists call Canada. On an entirely different note, Einstein and Bohm, despite their age difference, were known to go out on double dates. (The best-known story of their love lives involves two sisters.)

NIELS BOHR (1885–1962)

A family man, happily married for 50 years to Margrethe. They had six sons. One died in childhood; one died in a boating accident; the other four went on to success in medicine, engineering, law and physics. The lawyer played hockey for Denmark. Bohr and his brother Harald were sportsmen, although Bohr as a goalie was once caught doing calculations on the goalpost, and the physicist, Aage, won the Nobel Prize in 1975.

Bohr led the group of physicists who developed one of the greatest theories of all time – at his home. Heisenberg, Schrödinger, Fermi and many others came to stay with him and his family in Copenhagen, wore down the floor at either end of the ping-pong table, wrote letters home, munched on sandwiches and ran around with his laughing children.

Bohr, who was a member of the Danish underground during the Second World War, had a portentous walk with his German friend and colleague Heisenberg in 1941. No one knows exactly what was said, except that Bohr was extremely shocked and the friendship with Heisenberg ended there and then.

Possibly Heisenberg told Bohr of the Germans' plans for an atom bomb. Within months Bohr had contacted the Americans and was whisked off to the USA. As a direct result of Bohr's concern, work on the atom bomb was stepped up at Los Alamos.

MBV Bohr had an unfortunate way of speaking. Einstein's comment about Bohr's lack of ability to allow words to come out of his mouth properly was: 'He utters his opinions like one perpetually groping and never like one who believes himself to be in possession of the definite truth.' A fellow physicist once drew a cartoon of Bohr talking to a friend who is bound and gagged. Bohr is saying 'Please, please may I get a word in?'

MAX BORN (1882–1970)

Known as 'The Probability Man', the Polish-born Born was already pretty nifty with the old physics before he made his vast contribution to quantum theory by sorting out the mathematical formulations and practicalities of the ideas of Heisenberg and Schrödinger. You should say that it is his technique – the Born approximation – which is used by working physicists far more than the philosophical speculations of Heisenberg and Uncertainty.

You should also be scandalised that Born did not win the Nobel Prize until the day before he turned 72, and even then had to share it. It's worth mentioning that Einstein's famous statement about quantum mechanics – 'I cannot believe that God would play dice with the universe' – was made in a letter to his great friend Born.

MBV Born was Olivia Newton-John's grandfather.

LOUIS-VICTOR DE BROGLIE (1892–1987)

De Broglie (pronounced de Broy) devised a simple mathematical relationship connecting the wave and particle properties of matter.

Einstein had proposed the possibility that there were such things as matter waves in 1905, and in fact it was Einstein's enthusiastic response to de Broglie's thesis that made his name – another instance of Einstein's support for quantum mechanics, despite his loathing of the claims that were made for it.

MBV De Broglie's great-great-grandfather was a French aristocrat who died on the guillotine during the French Revolution's 'Reign of Terror'.

PAUL ADRIEN MAURICE DIRAC (1902–1984)

The word associated with Dirac is 'antimatter', which he was the first to suggest might exist. Of Anglo-Swiss parentage and born in Bristol, he was taciturn except when he talked of beauty. As a child his natural shyness and reticence wasn't helped by his father's insistence that he speak French even though they lived in England. Dirac was Feynman's hero. He makes many physicists quite misty-eyed because he believed that calculations and theories should be endowed with beauty. His approach to mathematics was always elegant and simple. So was his approach to life.

MBV Dirac hated wasting words. Once asked if he took sugar, he answered 'Yes' and was surprised to be asked: 'How many lumps?' If he had wanted more than one lump, he as a mathematician would have specified the number.

FREEMAN JOHN DYSON (1923–)

Nobel-laureate Steven Weinberg has said that the Nobel committee 'fleeced' Dyson, who has won numerous scientific awards but has never won a Nobel Prize – although if there were prizes for infectious chortling he would be a shoo-in. Dyson was the first person to appreciate the power of Feynman diagrams – apart from Feynman himself, of course – and to present Feynman's work in a form that could be understood by other physicists. Dyson received a lifetime appointment at the Institute for Advanced Study in Princeton along with heartfelt thanks from J Robert Oppenheimer 'for proving me wrong'.

Two of Dyson's six children, Esther and George, write with intelligence and perception about modern technology, and Dyson himself has never ceased to comment on atoms, bytes and cells. Many see him as a climate change denier. Some say he simply insists on proper evidence-based arguments, as any hard-nosed scientist should.

MBV People notice that Dyson, an Englishman who has lived in the USA for over 50 years, speaks with a slight German accent. His wife is German.

PAUL EHRENFEST (1880–1933)

Einstein's great friend, Ehrenfest had tears rolling down his cheeks during Einstein's debate with Bohr. He forcefully reminded Einstein not to be as rigidly against quantum mechanics as people had been against relativity. Yet he himself called the quantum theorists 'Klugscheisser' – clever shit.

Many have felt overwhelmed by quantum mechanics, but Ehrenfest's was an extreme case.

MBV The letter he sent before he committed suicide began: 'My dear friends Bohr, Einstein [and others]. In recent years it has become ever more difficult for me to follow developments [in physics] with understanding. After trying, ever more enervated and torn, I have finally given up in DESPERATION.'

ALBERT EINSTEIN (1879–1955)

Einstein was married twice (to Mileva and to his cousin, Elsa). His first child with Mileva was born before their marriage, and had to be given away because they were too poor to provide for it. His second son spent most of his life in a mental hospital.

Einstein worked alone. People who knew him well said that he was somehow always 'apart'. He said: 'I am not much good with people...I feel the insignificance of the individual and it makes me happy.' It is not surprising that he kept insisting that subatomic particles had to be thought of as separate entities – 'discrete'.

MBV Einstein died in the USA, far from his place of birth in Germany. Reporters jostling around the nurse who had been the only person present when he died were told that, yes, the great sage had spoken as he took his last breath. What did he say? 'I'm sorry,' said the nurse, 'I don't speak German.'

ENRICO FERMI (1901–1954)

Born in Rome, Fermi was rare in that he actually did his own experimental dirty work, even in the USA. The Fermi Award for innovation is still much coveted by physicists. Fermi built the first nuclear reactor.

MBV When he established the first chain reaction, Fermi made a phone call to a military colleague, saying calmly and in code, 'The Italian navigator has landed in the new world.' New world indeed.

RICHARD FEYNMAN (1918–1988)

Feynman's work on light and matter perfected what is possibly the most accurate theory ever developed in science. His famous diagrams show quantum interaction. You should always stress that although there is so much probability in quantum mechanics, there is also accuracy and precision.

Feynman (it rhymes with lineman) was a hugely popular figure, as much for his personality as for his mathematical wizardry. His diagrams were controversial because no one knew how he managed to derive them – or visualise them – but even physicists are unwilling to argue for long about a drawing that saves hundreds of pages of algebra.

MBV A brilliant lecturer, Feynman enjoyed doing sums

on paper napkins at nightclubs and was an accomplished picker of locks which, if you are dealing with atomic secrets and one of your best friends is the spy Klaus Fuchs (who gave US atomic secrets to the USSR), is a pretty risky business. Feynman had his kookier moments, such as his refusal, for some months, to brush his teeth or to wash his hands after urinating because of his ideas about germs. Late in life he became famous all over again for explaining why the *Challenger* space shuttle disaster happened: sealing rings became less resilient and were subject to failures at ice-cold temperatures in space.

MURRAY GELL-MANN (1929–)

Gell-Mann had an office across the way from Feynman at the California Institute of Technology in Pasadena in the 1950s, and was judged the more urbane of the two. He sprinkled his scientific work with literary and classical references, and ordered particles into arrangements or families which he called the eightfold way in honour of the Buddha.

Gell-Mann himself is not known for his tranquillity. Indeed, he is a renowned curmudgeon. He hasn't even begun to consider the notion that fools could be suffered and would find it unthinkable that anyone might do so gladly.

MBV Gell-Mann's most famous contribution is the 'quark' which he both discovered and named. Late in life Gell-Mann has been devoting his time to the origins of human languages.

STEPHEN HAWKING (1942–2018)

Hawking (together with Sir Roger Penrose) proved that the beginning of the universe was a 'singularity', a mathematical point of infinite density, the explosion of which was the Big Bang. Hawking fathered a subject whereby, instead of wrestling to bring all the forces together, the quantum-mechanical implications for gravity alone are studied. It's called quantum cosmology.

MBV Hawking loathed the idea of parallel universes. He once said, 'When I hear the words "Schrödinger's cat", I reach for my gun.'

Hawking's ashes are to be interred in Westminster Abbey near the graves of Ernest Rutherford, Charles Darwin, J.J. Thomson and Isaac Newton.

WERNER HEISENBERG (1901–1976)

Famous for his uncertainty principle (published around his 26th birthday), Heisenberg is also notorious for having gained some advancement during the Hitler years in Germany. But no one quite understands what his role was because plans for a German atom bomb were abandoned by order of the Führer.

MBV Heisenberg's musings about the observer and the observed have been exploited far beyond his intentions, and you should scoff at the cavalier use of the word 'uncertainty', which is not an excuse to avoid choosing a career, proposing marriage or crossing a street.

PETER WARE HIGGS (1929–)

It was with others that Higgs proposed the mechanism by which particles are endowed with mass by interacting with a field, which is carried by bosons. So it is truly remarkable – and an indication of the way science and our culture operates – that, famously, the mechanism is called the Higgs mechanism, the field is called the Higgs field, and the bosons are called Higgs bosons.

MBV His father was a sound man at the BBC.

LISE MEITNER (1878–1968)

Meitner was an Austrian, later Swedish, physicist who worked on radioactivity and nuclear physics. Meitner was part of the team that discovered nuclear fission, an achievement for which her colleague Otto Hahn was awarded the Nobel Prize. Meitner is often mentioned as one of the most glaring examples of women's scientific achievement overlooked by the Nobel committee.

MBV Meitner was taught by the legendary nineteenth-century physicist Ludwig Boltzmann and then by Max Planck, who until then had rejected any women wanting to attend his lectures. After one year, Meitner became Planck's assistant. Otto Hahn was also working for Planck.

WOLFGANG PAULI (1900–1958)

At the age of 20, Pauli wrote a 200-page encyclopaedia entry on the theory of relativity. It was his ideas that led to the discovery of the neutrino, but he is best known for his exclusion principle.

Pauli was excluded many times from bars for being pixilated. Even when sober, he had no trouble speaking his mind. To one student he said, 'Ach, so young and already you are unknown'; to another, 'That isn't even wrong.' He even put Einstein down for not seeing the difference between mathematics and physics. It was to Pauli that Bohr made the legendary remark: 'Your theory is crazy, but it's not crazy enough to be true.'

MBV Pauli, an Austrian, was also fascinated by subatomic particles and consciousness, collaborating for some time with psychologist Carl Jung, whose patient he was for a while. Their association may not have proved much, but it probably made them feel better.

Pauli, along with Feynman, had an obsession with the number 137, which is important in physics and in the ancient wisdom of the Kabbalah – the real thing (not Madonna's idea of Dolce and Kabbalah). The universe is 13.7 billion years old, the number 137 refers to the absorption of light by matter and – if you use a little mathematical magic – unites the electron charge, the speed of light and Planck's constant. In ancient tradition, 137 unites wisdom and prophecy as well as the spiritual and the material. When Pauli was taken ill in Zürich with pancreatic cancer in 1958, he was convinced his time had come when he was put in hospital room number 137. He was right.

> Planck could never come to terms with quantum mechanics, of which he himself was a pioneer.

MAX PLANCK (1858–1947)

Famous for Planck's constant: the energy of a light wave is always proportional to its frequency. The mathematically minded marvel at the strange relationship between Planck's constant and Heisenberg's uncertainty principle. By multiplying Heisenberg's two uncertainties you actually get Planck's constant. Truth is stranger than fiction.

Planck's constant is represented by h, which has the value of 6.63 times 10 to the power of minus 34 Js (joule second). It is always 6.63 x 10 to the power of minus 34 Js, and the energy of a photon is h – or 6.63 times 10 to the power of minus 34 Js multiplied by f, where f is the frequency of the wave. It should be clear by now why jokes about being thick as a Planck are not such a great idea.

He could never come to terms with quantum mechanics, of which he himself was a pioneer.

MBV Planck's son Erwin was executed in 1945 for attempting to assassinate Hitler.

ERNEST RUTHERFORD (1871–1937)

Rutherford was a New Zealand-born physicist who sang 'Onward Christian Soldiers' loudly and out of tune all day and every day for the whole of his life. His loudness was legendary and his own researchers manufactured an electrical sign in their lab which read, 'TALK SOFTLY PLEASE'. Rutherford set down the groundwork for the development of nuclear physics by discovering the alpha particle, the nucleus and the proton. He was also wise enough to employ Niels Bohr, as well as Hans Geiger, who developed the Geiger counter as a result of sitting in the dark and totting up flashes of radiation.

MBV Rutherford contributed to the atom bomb by speaking so loudly and firmly against it that he infuriated a young Hungarian physicist called Leó Szilárd. Within the decade Szilárd was a leader of the Manhattan Project.

Rutherford used to say, 'In science there is only physics; everything else is stamp collecting.' In 1908, when he won the Nobel Prize, it was for chemistry.

ERWIN SCHRÖDINGER (1887–1961)

A brilliant physicist who stands out because of his extraordinary intellectual versatility, Schrödinger's numerous contributions to science include an extremely useful wave equation, and a handy and profound book about quantum physics and genetic structure.

MBV In 1944 Schrödinger published a landmark book, *What Is Life?* Both James D Watson and Francis Crick, co-discoverers of the structure of DNA, credited Schrödinger's book with presenting an early theoretical description of how the storage of genetic information would work. The great physicist had become a pioneering molecular biologist.

Such a love of nature shouldn't lead you to think that, despite the fact that he is most famous for a dratted cat, he would actually own one.

The existence of antimatter is one of the arguments put forward for the existence of parallel universes. At least, in this universe it is.

WHAT IT ALL MEANS

Apart from the obvious fun that can be derived from believing impossible things for as much as half an hour a day, with six of them before breakfast, quantum mechanics is a big part of everyday life. It deals with the most basic of basic stuff. It is behind every chemical reaction, every biological and medical miracle. It underlies all of existence in some way.

And therein lies the controversy. For the holists, what is so powerful is the discovery of these fundamental interrelationships which permeate the entire universe and make it a cohesive (w)hole. They feel that Western civilisation should finally resign its obsession with dividing, compartmentalising and separating – an obsession that has governed intellectual activity since the ancient Greeks, and especially since Aristotle, whom they regard as the chief villain of the piece.

They say it's time to stop dismembering and begin remembering. They hold divisive thinking responsible for all the world's splitting, clefting and rending; they say it's what causes revolutions, riots and wars.

The holists latch on to statements like that made by John Wheeler, an American physics professor who worked with some of the greats, including Bohr. Wheeler stated: 'Nothing is more important about quantum physics than this: it has destroyed the concept of the world as "sitting out there".' Wheeler was very firm: this is no longer 'a universe for observation'. It is a universe that requires, demands and expects participation.

Reductionists don't buy that. Imagine a performer in a theatre asking if anyone hates audience participation. A reductionist would fall into the trap and say, 'Yes! Me!'

THE CUTTING EDGE

There are two fashionable topics at the cutting edge, or even the outer limits, whose popularity has mushroomed as a result of quantum mechanics. The first is mysticism, the second is consciousness.

1. Mysticism

To reductionists, this means anything from wearing a saffron robe to having a fondness for breathing too deeply. To them there is no worse word.

Supreme caution must be exercised with mysticism in the presence of any member of the scientific establishment, unless you are tired of life.

2. Consciousness

The fact that it is considered to be no more than a possibility that consciousness is connected with phenomena in the quantum realm should not prevent you from pontificating about it. Sir Roger Penrose, sometime colleague of Stephen Hawking, goes farther. For him, it's a 'definite possibility'.

Evidence for such a convinced 'maybe' lies in the swiftness with which the human brain makes choices. For instance, sifting all the visual alternatives as the eye focuses on a single printed word takes the brain less than a tenth of a second; a standard computer would have to start calculating before time and the universe began, so you wouldn't want to be stuck behind one of them on a roundabout.

To reductionists, mysticism means anything from wearing a saffron robe to having a fondness for breathing too deeply.

But speculation about the act of observation takes the issue far beyond the ordinary miracle of the brain's astounding speed, beyond settling the fate of Schrödinger's cat. It takes it towards affecting reality itself, and even into the clutches of Bishop Berkeley's assertion in the eighteenth century that mind creates matter.

With the double-slit experiment, are the particles and waves merely sensitive to screens and other equipment? Or are they picking up messages from the physicist's brain?

John von Neumann, biologist George Wald and physicists David Bohm and Arthur Eddington all made statements to the effect that the universe is mind-stuff, but it is difficult to get out of one's head for long enough to prove it, even in this day and age. What's more, no scientist would risk making statements to that effect in an actual paper.

More and more, though, in popular science magazines, writers are venturing to suggest that a watched pot might well boil – and boil faster – if your mind asks it to, while the fridge defrosts, the furniture rearranges itself and the loft gets converted.

The softer option is to say that particles of matter and particles of mind may come into being together, but that in any case our self-awareness and what seems to be some kind of consciousness at the quantum level are in profound communication.

Consciousness is also non-local. No one can say where the mind is, or how far the effects of thought can reach. The anthropic principles (both strong and weak – *see* 'Glossary') suggest that in Homo sapiens there are intellectual capacities that are there for a reason, that somehow human beings are obliged to help the universe through the next stage – although it is unclear whether this means giving it homework, or getting stoned with it and talking all night.

You can be supremely engaging and effective, without being too technical, by telling people about the messenger particles in the brain and how profoundly they reinforce the interconnectedness of everything.

Messenger particles – which by now you will be able

to call 'bosons' without flinching – are the strongest, most basic forces of relationship in nature. If anything relates more fervently, it hasn't yet been found. Indeed, within the brain, all the forces are active, including gravity, which is nothing less than the force that holds the entire universe together.

Perhaps there are such things as thought particles. And perhaps they know that this is being supposed. Perhaps they supposed it first. Perhaps the supposing happened simultaneously. Just a thought.

TO BE CONTINUED...

If questions in the quantum realm go to the root of everything and the answers that physicists seek will reveal the secret to Life, the Universe and Everything, then it is easy to see why, whenever you raise the subject, you are firing the starting gun for a conversational marathon.

The quantum universe has different rules about endings and beginnings. Many physicists believe that the Big Bang was a quantum event, the explosion of a single particle out of which the universe ballooned.

Whether the universe is contracting or expanding – crunching down to a single particle or creating from each particle a potential universe – there is nothing to stop you proposing that the universe, which is said to have begun with a Big Bang, could eventually evolve, dissolve and resolve into a Little Whimper. How could that ever happen?

The question doesn't apply.

There's no point in pretending that you know everything about the quantum universe – nobody does – but if you've got this far and absorbed at least a modicum of the information and advice contained within these pages, then you will almost certainly know more than 99% of the rest of the human race about what it is, why it is, how it is, where it is, and why nobody has much idea about what's going on.

What you now do with this information is up to you, but here's a suggestion: be confident about your newfound knowledge, see how far it takes you, but above all have fun using it. You are now a bona fide expert in the art of bluffing about a subject that is so impenetrably complicated that the world's finest minds are still confounded by what it all means.

And remember: when in doubt, simply answer: 'They're looking into it at Cern.'

GLOSSARY

Anthropic principle The theory that the universe has evolved so that human beings could come along. Because of this they have a purpose and their minds matter. The strong anthropic principle states that people matter a lot. The weak anthropic principle states that human beings matter quite a lot, especially if they understand the EPR paradox.
Antimatter Matter composed entirely of antiparticles.
Atom The smallest unit of any material which still retains the characteristics of that material. Obviously, if you go into the subatomic realm, anything can happen and probably will. The word 'atom' comes from the Greek meaning 'that which cannot be cut or split'. Right. And the *Titanic* was that which could not be sunk.
Big Bang Generally accepted notion that time had a beginning, and an incredible amount of action during the opening credits.
Billion Big number which should be bandied about with abandon. The British billion is traditionally a million million. In physics, the American billion is used: a thousand million. Piffling in either case.

Constant A comforting predictable factor. The speed of light is a constant. So is the constant discovered by Max Planck. Useful in science and in life: buses are always late, they don't write songs like that any more, and money is never going to grow on trees.

Dark energy A purely hypothetical form of energy. It is said to permeate all of space. Dark energy was put forward as an idea in the 1990s, when it seemed as if the expansion of the universe was not slowing because of gravity but accelerating because of, well, something. It has been ventured that roughly 68% of the universe is made up of it.

Dark matter A purely hypothetical form of matter. Makes up about 27% of the universe. Add that to the 68% of the universe that is dark energy and you will see that we understand less than 5% of the universe. That man Rumsfeld with his 'known knowns' and 'unknown unknowns' was onto something after all.

Electromagnetism A weak force. Ultraviolet rays from the sun are electromagnetic. So are X-rays, microwaves in ovens and radio waves.

Elementary particle A particle without any internal structure which is therefore considered to be at the very basis of matter. Not by any means a building block.
In other words, the smallest thing there is. So far.

Energy Oomph, pizazz, vigour. In physics, as in life, it also means the 'capacity to do the work'.

Gluons Mega-strong messenger particles which bind everything together in the nucleus. Far more binding than marriage vows, contracts or even eggs.

God Among physicists, even though they work at the

boundaries of the cosmos, please, please don't mention God. If God were actually to arrive during one of your discussions, berate Him for providing so little evidence.

Gravity The binding force of the universe, central to relativity, left out of the quantum dance. If anyone doubts gravity's existence, point to the ageing process.

GUT Grand unifying theory. A theory of everything which unites all the forces, the search for which goes on and on. On that, at least, physicists are united.

Laser light amplification by stimulated emission of radiation The emission is stimulated from an excited source. (Don't even think of making jokes about this.)

Mechanics Branch of applied maths dealing with motion and tendencies to motion. (Someone's got to do it.)

Molecule Two or more atoms in a 'bound system'. There are 92 naturally occurring substances (called elements) and another 17 man-made ones (so far), the atoms of which can combine to form molecules which ultimately form everything from guacamole to Guatemala.

Nucleus The dense core of an atom, made up of protons and neutrons. Nuclear energy and the nuclear bomb refer to this kind of nucleus. The nuclear family is so called because it, too, is a basic unit; to point out that families split, experience fallout or blow up is in very bad taste.

Particle Small part (sometimes the least possible part) of something. It is the central mystery of quantum mechanics that a particle sometimes behaves like a wave (*see* overleaf).

Probability Bluffers need to know that a number of

physicists prefer notions of probability to the use of the word 'uncertainty'. Physics graduates are much sought-after for jobs in the financial markets because they know about probabilities, waywardness and wildness from quantum mechanics. These prized whizz-kids are called 'quants'.

Quantum If you don't know now, you never will. But that doesn't mean you won't have a happy life.

Superstring theory Theory that minuscule vibrating strings will be, for certain, the smallest possible anythings in nature. A neat solution to bring the threads together? Reductionists are already sharpening their scissors.

Time Still not found in the quantum realm. Past and future particles just go jiggling and jiggling and jiggling along.

Wave A 'disturbance' like a ripple, a shiver or a vibration. It should be obvious to anyone – other than a physicist – that of course a tiny particle could somehow connect to the outer reaches of the universe.

BLUFFER'S NOTES